Manual on Postharvest Handling of Mediterranean Tree Fruits and Nuts

Manual on Postharvest Handling of Mediterranean Tree Fruits and Nuts

Edited by

Carlos H. Crisosto

University of California, Davis, USA

and

Gayle M. Crisosto

University of California, Davis, USA

CABI is a trading name of CAB International

CABI	CABI
Nosworthy Way	WeWork
Wallingford	One Lincoln St
Oxfordshire OX10 8DE	24th Floor
UK	Boston, MA 02111
	USA

Tel: +44 (0)1491 832111
Fax: +44 (0)1491 833508   T: +1 (617)682-9015
E-mail: info@cabi.org   E-mail: cabi-nao@cabi.org
Website: www.cabi.org

A catalogue record for this book is available from the British Library, London, UK.

Library of Congress Cataloging-in-Publication Data
Names: Crisosto, Carlos H., editor. | Crisosto, Gayle M., editor.
Title: Manual on postharvest handling of Mediterranean tree fruits and nuts / edited by Carlos H. Crisosto and Gayle M. Crisosto.
Description: Oxfordshire ; Boston, MA : CABI, [2020] | Includes bibliographical references and index.
Identifiers: LCCN 2020018092 (print) | LCCN 2020018093 (ebook) | ISBN 9781789247190 (epub) | ISBN 9781789247183 (ebook) | ISBN 9781789247176 (paperback)
Subjects: LCSH: Fruit–Handling–Mediterranean Region–Handbooks, manuals, etc. | Nuts–Handling–Mediterranean Region–Handbooks, manuals, etc. | Fruit–Postharvest technology–Mediterranean Region–Handbooks, manuals, etc. | Nuts–Postharvest technology–Mediterranean Region–Handbooks, manuals, etc.
Classification: LCC SB360 (ebook) | LCC SB360 .M425 2020 (print) | DDC 634/.046–dc23
LC record available at https://lccn.loc.gov/2020018092

References to Internet websites (URLs) were accurate at the time of writing.

ISBN-13: 9781789247176   (paperback)
          9781789247183   (ePub)
          9781789247190   (ePDF)

Commissioning Editor: Rachael Russell
Editorial Assistant: Emma McCann
Production Editor: Shankari Wilford

Typeset by Exeter Premedia Services Pvt Ltd, Chennai, India
Printed and bound in the UK by Severn, Gloucester

Contents

Contributors

Besada, Cristina, Instituto Valenciano de Investigaciones Agrarias (IVIA), Ctra. Moncada-Naquera, Km. 4, Moncada, Valencia 46113, Spain. Email: besada_cri@gva.es

Campos-Vargas, Reinaldo, Centro de Biotecnología Vegetal, Facultad de Ciencias Biológicas, Universidad Andrés Bello, República 217, Santiago, Chile. Email: reinaldocampos@unab.cl

Colelli, Giancarlo, Dipartimento di Scienze Agrarie, degli Alimenti, e dell'Ambiente, University of Foggia, Foggia, Italy. Email: giancarlo.colelli@unifg.it

Connell, Joseph H., University of California (UC) Cooperative Extension Butte County, 2279-B Del Oro Avenue, Oroville, CA 95965, USA. Email: jhconnell@ucanr.edu

Crisosto, Carlos H., Department of Plant Sciences, One Shields Avenue, University of California, Davis, CA 95616, USA. Email: chcrisosto@ucdavis.edu

Echeverría, Gemma, IRTA (Institut de Recerca i Tecnologia Agroalimentaries), Parc Científic i Tecnològic Agroalimentari de Lleida, Parc de Gardeny, Edifici Fruitcentre, E-25003, Lleida, Spain. Email: gemma.echeverria@irta.cat

Ferguson, Louise, Department of Plant Sciences, One Shields Avenue, University of California. Davis, CA 95616, USA. Email: lferguson@ucdavis.edu

Lichter, Amnon, Department of Postharvest Science, Agricultural Research Organization – The Volcani Center, 68 HaMaccabim Road , POB 15159, Rishon LeZion 7505101, Israel. Email: vtlicht@agri.gov.il

López Corrales, Margarita, Centro de Investigación Finca la Orden-Valdesequera, Guadajira-Badajoz, Spain. Email: margarita.lopez@juntaex.es

Manganaris, George A., Department of Agricultural Sciences, Biotechnology & Food Science, Cyprus University of Technology, PO Box 50329, 3603 Lemesos, Cyprus. Email: george.manganaris@cut.ac.cy

Saa Silva, Sebastian, Almond Board of California, 1150 Ninth Street, Suite 1500, Modesto, CA 95354, USA. Email: ssaa@almondboard.com

Salvador, Alejandra, Instituto Valenciano de Investigaciones Agrarias (IVIA), Ctra. Moncada-Naquera, Km. 4.5, Moncada, Valencia 46113, Spain. Email: asalvado@ivia.es

Serradilla, Manuel Joaquin, Centro de Investigaciones Científicas y Tecnológicas de Extremadura, Instituto Tecnológico Agroalimentario de Extremadura, Badajoz, Spain. Email: manuel.serradilla@juntaex.es

Valero Garrido, Daniel, Deptamento de Tecnología Agroalimentaria, División Tecnología de Alimentos, Universidad Miguel Hernández, Carretera de Beniel, Km. 3.2, Orihuela, Spain. Email: daniel.valero@umh.es

Preface

For over 30 years, I have been a University of California (UC), Davis, California State-Wide Pomology Postharvest Specialist in charge of fresh fruit and nuts. Prior to that, in the early 1980s, I was an Assistant Pomology Professor at the Universidad Católica de Chile, working with deciduous fruit. Then, for less than 2 years, as a postdoc I investigated tropical fruit postharvest physiology, including coffee and sugarcane, at the Hawaiian Sugar Planters' Association (Oahu and Kauai).

In my current position at UC Davis that I started in 1990, the focus of my research and extension programs is postharvest biology and technology of peaches, nectarines, plums, and apricots (stone fruit). The main objective of my research program is to better understand the orchard and postharvest factors that control fruit flavor and postharvest life and to develop and implement technologies to overcome fruit industry problems. Currently, I am applying genomic techniques to identify gene(s) responsible for fruit sensory attributes (both desirable and undesirable) and investigating physiological disorders such as chilling injury, tissue browning and "off-flavors".

In the early 1990s, I added additional crops to my areas of interest: table grapes, kiwifruits, persimmons, fresh figs, pomegranates and nuts. I was living in the San Joaquin Valley and my stone fruit clientele also grew, packed and marketed these commodities. I enjoyed interacting with industry people and especially with my peers: UC Pomology Specialist Mr. F. Gordon Mitchell, University of California Cooperative Extension (UCCE) Kern County farm advisor Mr. Don Luvisi, and UCCE Tulare County farm advisor Mr. Kevin R. Day, Professor DeJong, and Themis Michailides. Some of these Mediterranean tree fruits are relatively minor crops in California with limited support, but are very important in other countries, which dedicate intensive economic resources to solving their postharvest problems. It can be difficult to access information on some of these Mediterranean tree fruits. The information may not be available, not distributed or is in other languages and most results of new research studies have not been made public.

During my career, I have acquired much information and knowledge from discussing tree fruit and nut handling with my non-Californian peers at international meetings. I facilitated hosting visitors and students at the Kearney Agricultural Center and at Davis campus and established collaboration on common postharvest problems. This has resulted in working relationships and friendships that still remain active.

Production and marketing of these Mediterranean tree fruits supports our communities and provides flavorful and nutritious food to consumers. Recent medical studies using human panels support the benefits of consuming Mediterranean fruits as part of a healthy diet to promote a longer and more active life. Thus, I envisioned the creation of this *Manual on Postharvest Handling of Mediterranean Tree Fruits and Nuts*, intended to include the most important Mediterranean tree fruits that contribute to the world's agricultural economy.

This manual covers relevant postharvest topics, with an emphasis on knowledge useful to solving current industry problems worldwide. In California, members of the UC Davis-UC Agriculture and Natural Resources (ANR) system have developed detailed information on almonds, table grapes, peaches, and pistachios over the years, but we lack detailed current information on peaches, pomegranates, and particularly, fresh figs and persimmons. For this reason, I invited peers from all over the world with expertise from active research programs and years of experience working with these Mediterranean tree fruits to collaborate on or coauthor the chapters. Thus, I am blending contributions and information generated by UC with that of other key groups in Mediterranean areas.

With the contributions of our international and domestic peers and well-chosen pictures, all following a standard easy-to-follow format, this publication should provide relevant information with a high impact on tree fruit industries, and students worldwide. The Appendix provides a brief summary of the storage requirements and benefits of postharvest treatment for Mediterranean tree fruits.

Carlos H. Crisosto

Acknowledgments

The authors greatly appreciate the time and thought given by others reviewing this book and the many valuable suggestions that have been contributed. Thanks are due to growers and packers who opened their operations and allowed me to take photos to complement the text in this book.

Special thanks to Mary Lou Mendum and Andrea Velasquez for their editing, formatting, comments, and careful review.

Dedication

This book is dedicated to the United States Land Grant System, the Department of Pomology, UC Davis and the UC ANR Research Stations: the Kearney Agricultural Center and the many people who made them. Below is a brief history of these organizations.

US Land Grant System

The land grant system began in 1862 with a piece of legislation known as the Morrill Act. This law gave states public lands, provided the lands were sold or used for profit and the proceeds used to establish at least one college—hence, land grant colleges—that would teach agriculture and the mechanical arts. Land grants for the establishment of colleges of agriculture and mechanical arts were later also given to US territories and the District of Columbia. The legislative mandate for these land grant colleges helped extend higher education to broad segments of the US population.

The Department of Pomology, UC Davis

The Department of Pomology was devoted solely to fruit production and postharvest science. It was responsible for instruction, research and continuing education related to the biology and production of fruit and nut crops, maintenance of their postharvest quality and uses for their many products. It provided world leadership in the science and technology of production and handling of fruit and nut crops. Over the years, many members have received national or international recognition for their research, teaching or extension activities and several serve in leadership roles in academic or industry organizations.

The UC Davis Department of Pomology has historic roots that trace back to the beginnings of the University of California. Public interest in agricultural instruction and research encouraged the California legislature to create the University of California in 1868, with the first campus located at Berkeley. In 1905, the California University Farm Bill authorized the

University of California Regents to purchase land for the establishment of a university farm school. In 1906, Davisville was selected as the site for the university farm and a year after, the university farm was dedicated and Davisville was renamed Davis. By 1909, the University Farm School had opened at Davis and the UC Davis Department of Pomology began during 1912–1913 as the Division of Pomology, located on the Berkeley campus after a reorganization of the Department of Horticulture there.

Research in pruning, irrigation, soil fertility, pollination, and plant nutrition produced historically relevant results that were of direct, practical benefit to the California fruit industry in the early years of the Division of Pomology's founding. The work also contributed meaningfully to scientific understanding of pollination and fertilization processes in fruit set and fruit development. Study of fruit tree rootstocks at Davis began when rootstock material was incorporated with the student laboratory planting and the variety collections. One of the more important areas of research begun in the first decade of the Division of Pomology was that of postharvest handling, storage and shipping of fruit. Over the years, this research has contributed to effective marketing of California fruits and fruit products across the country and around the world, helping to assure the success of the state's fruit industry.

In 1953, the Division of Pomology became the Department of Pomology, gaining independence from the Department of Horticulture just as the University of California, Davis, would gain its independence in 1959, when it was designated an independent campus of the University of California. The Department of Pomology continued its previous studies on fruit variety improvement, soil and water management, fruit thinning, and fruit harvesting, handling and storage. Cooperative work with the departments of Agricultural Engineering and Food Science and Technology included research on fruit maturity and ripening, as they affected canning practices and the quality of the processed product. Fruit breeding programs made dramatic progress, especially with plums, peaches, and strawberries. Cooperative work extended beyond the university: during the late 1960s and early 1970s, the Department of Pomology played a significant role in a cooperative education and research program between the University of Chile and the University of California known as the "Convenio", designed particularly to assist the University of Chile to strengthen its teaching and research in agriculture, with the Department of Pomology assisting with fruit culture.

The UC Davis Department of Pomology is now part of the Department of Plant Sciences, which was created by merging the four commodity-based departments of Agronomy and Range Science, Pomology, Vegetable Crops, and Environmental Horticulture. This pooling of manpower and resources has fostered better focus on the teaching, research and outreach missions of one of the leading academic programs in agricultural and environmental sciences in the country.

UC ANR Research Stations: the Kearney Agricultural Center (UC KAC)

The UC KAC, currently called UC KARE, is uniquely positioned to support extension of UC's research-based information to farmers and members of the public because of its locations, facilities and associated personnel. There are research orchards, vineyards and the F. Gordon Mitchell post-harvest laboratory on site and nearby commercial plantings of key specialty crops. Extension activities include meetings, field days, tours, web conferences, ANR publications, peer-reviewed journals, and contemporary and emerging electronic tools such as online learning, web content systems and repositories, social media, impact and evaluation tools, and specialized and public media outlets. Programs are offered and carried out in collaboration with county- and campus-based UC academics, federal, state and local government, regulatory agencies, grower and commodity groups and private industry. Key clientele include almond, citrus, fig, kiwifruit, table grape, pomegranate, pistachio, stone fruit, and walnut growers, pest control advisors, spray technology industries and commodity groups, conservation and environmental groups and public agencies.

Carlos H. Crisosto

Conversion Tables

Table A. Conversion of metric system to US customary system.

Metric system			US customary system		
Linear measure			**Linear measure**		
10 millimeters	=	1 centimeter	12 inches	=	1 foot
10 centimeters	=	1 decimeter	3 feet	=	1 yard
10 decimeters	=	1 meter	5½ yards	=	1 rod
10 meters	=	1 dekameter	40 rods	=	1 furlong
10 dekameters	=	1 hectometer	6 furlongs	=	1 mile
10 hectometers	=	1 kilometer	3 land miles	=	1 league
Area measure			**Area measure**		
100 sq. millimeters	=	1 sq. centimeter	144 sq. inches	=	1 sq. foot
10,000 sq. centimeters	=	1 sq. meter	9 sq. feet	=	1 sq. yard
1,000,000 sq. millimeters	=	1 sq. meter	30¼ sq. yards	=	1 sq. rod
100 sq. meters	=	1 acre (a)	160 sq. rods	=	1 acre
100 acres	=	1 hectare (ha)	640 acres	=	1 sq. mile
100 hectares	=	1 sq. kilometer	1 sq. mile	=	1 section
1,000,000 sq. meters	=	1 sq. kilometer	6 sections	=	1 township
Volume measure			**Liquid measure**		
1 liter	=	0.001 cubic meter	4 gills (2 cups)	=	1 pint
10 milliliters	=	1 centiliter	2 pints	=	1 quart
10 centiliters	=	1 deciliter	4 quarts	=	1 gallon
10 deciliters	=	1 liter	**Dry measure**		
10 liters	=	1 dekaliter	2 pints	=	1 quart

Continued

Metric system			US customary system		
10 dekaliters	=	1 hectoliter	8 quarts	=	1 peck
10 hectoliters	=	1 kiloliter	4 pecks	=	1 bushel
Weight			**Weight**		
10 milligrams	=	1 centigram	27 11/32 grains	=	1 dram
10 centigrams	=	1 decigram	16 drams	=	1 ounce
10 decigrams	=	1 gram	16 ounces	=	1 pound
10 grams	=	1 dekagram	100 pounds	=	1 hundredweight
10 dekagrams	=	1 hectogram	20 hundredweight	=	1 ton
10 hectograms	=	1 kilogram			
1000 kilograms	=	1 metric ton			

sq., square.

Kitchen measurements.

3 tsp. = 1 tbsp.	$5\frac{1}{3}$ tbsp. = $\frac{1}{3}$ cup	2 cups = 1 pint	2 pints = 1 quart
4 tbsp. = ¼ cup	16 tbsp. = 1 cup	4 cups = 1 quart	4 quarts = 1 gallon

tbsp., tablespoon; tsp., teaspoon.

Table B. Conversion table.

To convert	Into	Multiply by	To convert	Into	Multiply by	To convert	Into	Multiply by
Centimeters	Inches	0.394	Kilogram	Grams	1000	Miles	Feet	5280
	Feet	0.0328		Ounces	35.274		Yards	1760
	Meters	0.01		Pounds	2.205		Kilometers	1.609
	Millimeters	10	Kilometers	Feet	3281	Ounces	Grams	28.35
Feet	Centimeters	30.48		Meters	1000		Pounds	0.0625
	Inches	12.00		Miles	0.621		Kilograms	0.028
	Meters	0.3048		Yards	1093.666	Pints	Liters	0.473
	Miles	0.0001894	Liters	Cups	4.226		Quarts	0.50
	Yards	0.333		Pints	2.113		Gallons	0.125
Gallons	Pints	8.0		Gallons	0.264	Pounds	Grams	453.59
	Liters	3.785		Milliliters	1000		Ounces	16.0
	Quarts	4.0		Quarts	1.057		Kilograms	0.454
Grams	Ounces	0.035	Meters	Centimeters	100	Quarts	Pints	2.0
	Pounds	0.002		Feet	3.281		Liters	0.946
	Kilograms	0.001		Inches	39.37		Gallons	0.25
Inches	Centimeters	2.54		Kilometers	0.001	Yards	Inches	36.00
	Feet	0.0833		Miles	0.0006214		Feet	3.00
	Meters	0.254		Millimeters	1000		Meters	0.914
	Yards	0.0278		Yards	1.094		Miles	0.0005682

Table C. Conversion of US customary units to SI units.

Customary unit	×	Factor	=	SI unit
ft	×	0.3048	=	M
ft^2	×	0.0929	=	m^2
ft^3	×	0.02832	=	m^3
ft^3/min	×	0.000472	=	m^3·sec^{-1}
ft^3/min/lb	×	0.00104	=	m^3·sec^{-1}·kg^{-1}
ft^3/min/lb	×	1.04	=	m3·sec^{-1}·tonne^{-1}
lb	×	0.4536	=	kg
ton	×	907.2	=	Mg (= 1000 kg = 1,000,000 g = 1 tonne)
lb/in^2	×	6.895	=	kPa
lb/in^2	×	51.713	=	mm Hg @ 0°C
lb/ft^3	×	16.02	=	kg·m^{-3} (kg·m^{-3} = g·l^{-1})
inches of water column	×	0.2487	=	kPa
Btu	×	1.055	=	kJ (= kW·sec)
Btu/lb/°F	×	4.1868	=	kJ·kg^{-1}·°C^{-1}
hp	×	0.746	=	kW
ton of refrigeration	×	3.52	=	kW of refrigeration
gpm	×	0.0631	=	l·sec^{-1}
gpm/ft^2	×	0.6791	=	l·sec^{-1}·m^{-2}
(°F-32)	×	5/9	=	°C

°F	°C	°F	°C	°F	°C
14	-10.0	51	10.6	88	31.1
15	-9.4	52	11.1	89	31.7
16	-8.9	53	11.7	90	32.2
17	-8.3	54	12.2	91	32.8
18	-7.8	55	12.8	92	33.3
19	-7.2	56	13.3	93	33.9
20	-6.7	57	13.9	94	34.4
21	-6.1	58	14.4	95	35.0
22	-5.6	59	15.0	96	35.6
23	-5.0	60	15.6	97	36.1
24	-4.4	61	16.1	98	36.7
25	-3.9	62	16.7	99	37.2
26	-3.3	63	17.2	100	37.8
27	-2.8	64	17.8	101	38.3
28	-2.2	65	18.3	102	38.9
29	-1.7	66	18.9	103	39.4
30	-1.1	67	19.4	104	40.0
31	-0.6	68	20.0	105	40.6
32	**0**	69	20.6	106	41.1
33	0.6	70	21.1	107	41.7
34	1.1	71	21.7	108	42.2
35	1.7	72	22.2	109	42.8
36	2.2	73	22.8	110	43.3
37	2.8	74	23.3	111	43.9
38	3.3	75	23.9	112	44.4
39	3.9	76	24.4	113	45.0
40	4.4	77	25.0	114	45.6
41	5.0	78	25.6	115	46.1
42	5.6	79	26.1	116	46.7
43	6.1	80	26.7	117	47.2
44	6.7	81	27.2	118	47.8
45	7.2	82	27.8	119	48.3
46	7.8	83	28.3	120	48.9
47	8.3	84	28.9	121	49.4
48	8.9	85	29.4	122	50.0
49	9.4	86	30.0		
50	10.0	87	30.6		

Almond

Carlos H. Crisosto[1]*, Sebastian Saa Silva[2], and Joseph H. Connell[3]
[1]*University of California, Davis, California, USA*
[2]*Almond Board of California, Modesto, California, USA*
[3]*UC Cooperative Extension Butte County, Oroville, California, USA*

Scientific Name, Origin and Current Areas of Production

Almonds belong to the genus *Prunus*, which includes all stone fruits, and belongs to the *Rosaceae* family. While other *Prunus* species, like peach or cherry, are grown for their fruits' juicy flesh or mesocarp, almond is grown for its seeds and it is classified as a nut. The cultivated sweet almond is *Prunus dulcis* (Mill.) D.A. Webb, but the genus also includes many wild species. Although similar, *Prunus amygdalus* is a bitter almond. The cultivated almond tree grown today originated from wild species in the deserts and foothills of Central and South-west Asia. By selecting and cultivating the sweet kernel specimens, their use was widespread in the ancient world perhaps as early as 4000 BC. Almonds have been cultivated for over 4000 years and starting about 450 BC were cultivated around the Mediterranean coastline from Turkey to Tunisia. Almonds were first introduced to California with the founding of the Spanish California missions in the late 1700s, but the large commercial industry was built with local seedling selections from varieties brought to California from the Languedoc area of southern France in the 1850s (Kester and Ross, 1996). The mild wet winters and hot dry summers of California's Mediterranean climate provided an environment in which almond trees could thrive in the Central Valley. California is the only US state that produces almonds commercially, producing the signature paper-shell variety 'Nonpareil', along with other soft-shell California types and a few hard-shell selections (Fig. 1.1). The soft-shell varieties are the basis of the California industry. Production under irrigated conditions in California accounts for 80% of the world crop.

*Corresponding author: chcrisosto@ucdavis.edu

© CAB International 2020. *Manual on Postharvest Handling of Mediterranean Tree Fruits and Nuts* (eds Carlos H. Crisosto and Gayle M. Crisosto)

Fig. 1.1. Almond Cultivars. Figure courtesy of Dr. Sebastian Saa, Almond Board of California, Modesto, California, USA.

Fruit Physiological Characteristics

The almond fruit is a drupe characterized by an outer fibrous layer or hull equivalent to the flesh of the stone fruits. The almond's hairy epidermis, or exocarp, the hull, is made of the pericarp and the mesocarp, and the shell or endocarp all derived from the ovary wall. The shell contains the seed or kernel, which is the primary commercial part of the fruit. Within the ovary, the ovules are enclosed by two layers called integuments, which eventually form the seedcoat, also called skin or pellicle. The ovule becomes the seed or kernel containing the embryo resulting from fertilization that will grow into the edible part of the future nut. Harvesting is usually carried out once the hull on all nuts is beginning to dehisce and the shell is exposed. The leathery hull is a by-product used mainly as feed for dairy cattle. The usual fruit weight in almond cultivars ranges from 8 g to 20 g. The expanded base of the flower or ovary will develop into the entire fruit. The shell ranges from very soft paper shells to very hard stony shells, and their morphology is variable between cultivars. The preference for each shell type depends on the growing conditions and the prevalent industry in the region. As for the fruit, the kernel or seed weight varies between cultivars, from 0.5 g to 1.5 g. The general trend in the industry is the preference for large kernels to improve yield and facilitate and cheapen the process of cracking and blanching. Almond flowers have a single carpel with two ovules, as in other stone fruits. The secondary ovule often degenerates, and a single kernel is produced. If the two ovules reach full development and are fertilized, double kernels are produced. The presence of double kernels is a cultivar trait. The edible kernel (primarily two cotyledons whose cells are filled with oil bodies and a small embryo) is surrounded by a shell and hull tissue. Almonds are relatively high in oil: 36–60% of kernel dry mass. Most of the fatty acids in almond oil are unsaturated, with the ratio of monounsaturated to polyunsaturated ranging from 2:1 to almost 5:1.

Ethylene production and sensitivity

Almonds produce very little ethylene and there are no documented responses to ethylene that might directly affect kernel quality (Kader, 1996).

Respiration rates

The low water content and/or water activity of properly stored kernels makes them relatively inert metabolically (King *et al.*, 1983; King and Schade, 1986). Respiratory rates are very low.

Chilling sensitivity

Almonds are not sensitive to chilling during storage.

Quality Characteristics and Criteria

Currently over 85% of the California almond crop is sold as shelled products but developing export markets include substantial interest in in-shell product. Recently, there were over 300 million pounds of in-shell almonds shipped to India and China. In-shell almonds should have shells that are uniform, with a bright color, and be free of adhering hull material, debris, signs of insect damage or decay. The shell should be intact and free of damage caused by the hulling operation, insects, or fungi. Kernels should be fully formed rather than shriveled and larger sizes are preferred. The "skin" of the kernel (pellicle) should be unbroken (free of damage caused during shelling or by insects or pathogens) and of uniform light brown color. Double, split, or broken kernels are negative factors. A complete description of US Federal quality standards can be found at https://www.ams.usda.gov/grades-standards/almonds-shell-grades-and-standards (USDA Marketing Service, 2019). Almond flavor should display a combination of sweet and oily aroma and absence of stale or rancid flavors. Optimal kernel texture is from crisp to chewy. Kernels should have <5–6% moisture, but kernels with <4.0% moisture tend to be brittle and hard. Almonds are one of the highest dietary sources of vitamin E, magnesium, and manganese; as well as an important plant-based source of vital minerals like calcium and potassium. Among nuts, almonds are a good source of fiber, protein, copper, phosphorous, riboflavin, and niacin (USDA Agricultural Research Service, 2019; https://fdc.nal.usda.gov/fdc-app.html#/food-details/170567/nutrients). Almonds contain 40–60% fats by weight and less than 10% is water (Sathe *et al.*, 2008). The two most abundant unsaturated fatty acids are oleic acid (18:1, 62–80%) and linoleic acid (18:2, 10–18%) in addition to a high concentration of "good" phenolics and tocopherols (~24 mg g^{-1}).

Sensory attributes (texture, taste, and aroma) and chemical characteristics (fats, antioxidants, and sugars) have been described for in-shell, raw, roasted, and blanched nuts for a large group of almond genotypes at harvest and storage (Franklin *et al.*, 2017, 2018; Franklin and Mitchell, 2019). The sensory profile including 16 attributes was measured by a trained panel as a sensory baseline to quantify sensory changes triggered by postharvest handling, drying, roasting, storage conditions, and other treatments. Most flavor attributes either increased or decreased with time; intensity of Clean Nutty aroma and Clean Nutty flavor associated with fresh almond (correlation value with respect to time (rT): –0.89 and –0.95, respectively) and Clean Roasted aroma and flavor (rT: –0.71 and –0.80, respectively) decreased with storage in both light roasted and dark roasted almonds (Franklin *et al.*, 2018). Sensory attributes related to oxidative rancidity such as Cardboard/Painty/Solvent, Soapy, and total oxidized increased in intensity over time (Franklin *et al.*, 2018), thus, total oxidized aroma and total oxidized flavor (rT: 0.91 and 0.95, respectively), as well as the oxidation-specific flavor attributes Cardboard (rT 0.86), Painty/Solvent (rT 0.96), and Soapy (rT 0.98). The mouthfeel attributes Pungent/Irritation/Burning (rT 0.94) and Astringent (rT 0.36) also increased over time, to a lesser extent. At the same time, consumer liking (acceptance, hedonic analysis) was determined using a large group of untrained consumers, thus, changes in consumer acceptance was related to specific chemical and sensory measurements allowing the creation of some market life prediction models (Cheely *et al.*, 2018). A number of volatile predictors of consumer liking were identified, including 2,5-dimethylpyrazine and 2- and 3-methylbutanal, which were predictors of the desired "Clean Nutty" and "Clean Roasted" attributes. Additionally, a number of volatiles correlated with rancid flavor attributes were identified, which may be used to predict rancidity in roasted almonds (Franklin *et al.*, 2018). Among them, hexanal,

Fig. 1.2. Navel orangeworm (NOW) kernel damage. Photo courtesy of Dr. Carlos H. Crisosto.

the most important predictor of total oxidized aroma, and heptanal and octanal were better predictors of average consumer liking and may be more reliable indicators of consumer perception of rancidity in roasted almonds.

Horticultural Maturity Indices

A primary incentive for rapid harvest of soft-shelled cultivars in California is to avoid costly navel orangeworm (NOW, *Amyelois transitella*) damage to almond kernels (Fig. 1.2). Beginning with a timely 'Nonpareil' harvest helps avoid early fall rains that delay harvest and decrease quality by increasing both worm damage and mold. The percentage of hull split correlates with nut removal by shaking, providing a field guide to acceptable maturity. The dry weight and drying rate of almond kernels during harvest have been characterized. When nuts on the tree reach 100% hull split, stick-tight hulls are minimized and nut removal by shaking is maximized. Keeping these parameters in mind, harvest operations are timed to optimize kernel quality. Almond maturation can be monitored externally by evaluating the extent of hull dehiscence. When the two halves of the hull are fully open to expose the shell, hulls readily separate and moisture content is low enough that nuts can be picked up from the orchard floor in a few days. Yield is maximized because the kernel's dry weight is no longer increasing, and almond removal from the tree is close to 100%. Almond maturation on a given tree is not uniform; development tends to be most rapid on the south and south-western faces high in the tree canopy and slower in the lower interior. The California industry favors a timely (early) 'Nonpareil' harvest that helps avoid NOW egg-laying in split hulls. Thus, harvest is matched to the time when the last almond on the lower interior of a tree has begun to

Fig. 1.3. General view of an almond orchard prepared for shaking. Photo courtesy of Dr. Carlos H. Crisosto.

split. Nut removal is near maximum, as is kernel dry weight. Nuts harvested very early are greener, are not open to the shell in the lower interior tree canopy and will produce more sticktights (hulls shriveled around the in-shell nut). Since they are greener and have a higher water content these almonds must dry longer on the orchard floor for 1–2 weeks before being picked up and hulled (Connell *et al.*, 1989, 1996).

Harvesting (Shaking and Picking) and Handling

Orchard floor preparation for harvesting

Almond harvest typically begins in early to mid-August and continues until late September for roughly 6–8 weeks depending on cultivars. Very few California almond orchards are cultivated. Those located in the Central Valley from Bakersfield to Chico are on flat land that facilitates irrigation and mechanical harvesting operations. Typically, pre-emergent strip weed control is used down the tree rows to control winter annual weeds. Orchard middles are mowed in the spring and sprays of approved translocated herbicides are used to control summer annual weeds followed by a final mowing. The orchard floor is smooth, firm, and free of weeds as harvest approaches (Fig. 1.3). All California almond orchards are irrigated: some with sprinklers, most with microsprinklers or drip irrigation, and a few are still flood irrigated. With sprinklers or

Fig. 1.4. Almond canopy showing 100% of almonds at the hull-split stage. Photo courtesy of Dr. Carlos H. Crisosto.

flooding, a final irrigation before harvest is used to fully recharge the soil profile. This enables the orchard to go through a long, dry harvest period with minimum water stress (Connell *et al.*, 1996; Reil *et al.*, 1996). Usually, the last preharvest irrigation is timed around 1–2 weeks prior to the onset of mechanical shaking to remove in-hull almonds from the trees. The incidence of sticktights can increase when severe deficit irrigation is applied in between hull split and harvest. Thus, to reduce severe tree stress utilizing regulated deficit irrigation, tree stress levels should be kept less than –0.15 MPa with microsprinkler or drip irrigation. This is accomplished with additional irrigation close to the time of harvest of each cultivar so that sometimes irrigation takes place by cultivar row, or with additional supplemental irrigation often applied to the orchard between 'Nonpareil' harvest and harvest of the pollenizer.

Determining shaking date

Two separate processes signal the approach of almond nut maturity. The first is hull dehiscence, in which the hull splits along the suture line, gradually separates from the shell, and begins to dry. Harvesting usually starts once 100% hull dehiscence (Fig. 1.4) is reached; at this stage the shell is mostly visible, and hulls are open wide and drying in the upper tree canopy (Connell *et al.*, 1989; Reil *et al.*, 1996). In the lower interior canopy, hulls are green and the suture is split just enough to be able to see a small portion of the shell. The second is the formation of an abscission layer at the nut—peduncle connection, at approximately the same time as hull dehiscence. There are still fiber attachments that must be disconnected at the time of harvesting.

Fig. 1.5. Almond shaker. Photo courtesy of Dr. Carlos H. Crisosto.

Fig. 1.6. View of shaker attached to the almond trunk. Photo courtesy of Dr. Carlos H. Crisosto.

Shaking

After almond hulls split and the nuts begin to dry, they are shaken to the orchard floor with mechanical shakers (Fig. 1.5). If nut removal is good and there is no bark damage (Fig. 1.6), the shakers will continue to harvest the earliest maturing 'Nonpareil' cultivar. If results are not satisfactory, the shaker will stop, wait a few days, and try again. When

Fig. 1.7. Almonds drying on the ground. Photo courtesy of Dr. Carlos H. Crisosto.

100% of the nuts in the canopy have split hulls, nuts have reached their full potential for both size and removal by mechanical shaking. This relationship serves as a field guide to acceptable maturity of the 'Nonpareil' cultivar (Connell *et al.*, 1989, 1996). Once all nuts have reached the hull-split stage, sticktight hull incidence decreases to insignificant levels. Following fruit removal at harvest as nuts drop to the orchard floor, the peduncle remains attached to the spur. Complete dehiscence requires internal tree moisture because the sides of the hull must be turgid to separate properly. If the fruits are subject to moisture stress, hulls may not dehisce but instead tighten on the shell. On trees less than 15 years old, a single trunk shake is all that is usually required. Large, old trees may require shaking of two or three major limbs.

Drying on the ground

Once shaken off the tree, nuts continue to dry on the orchard floor until the hulls will snap when bent (Fig. 1.7). Water is present in plant tissues in three forms: (i) bound water, bound to other constituents by strong chemical forces; (ii) adsorbed water, held by molecular attraction to adsorbing substances; and (iii) absorbed water, held loosely in the extracellular spaces by the weak forces of capillary action. The absorbed and adsorbed water constitute the "free water," most of which is removed by drying. Bound water is not removed except at very high temperatures that also decompose some organic matter. In general, it takes much longer for hulls and kernels to dry on the tree than on the ground. Thus, most of the final drying of almonds occurs naturally while they are on the orchard floor. It may take 5–14 days depending on hull moisture at the time of shaking. Then, the sweeping operation begins. The sweeper blows the nuts out of the tree row into the opposite middle and sweeps them toward the center of the middle it is working in. Sometimes a small amount of hand raking is needed around tree trunks to recover nuts missed by the sweeper. After two passes down the middle of each row in opposite directions, the mechanical sweeper forms a row of nuts that can be picked up.

Picking

Almonds are picked up from the orchard floor as soon as they are dry to avoid exposure to adverse weather conditions, especially rain, and to minimize fungal infection and insect damage. Exposure of almonds to wet and hot conditions results in concealed damage (CD), an internal disorder characterized by rust-brown to black discoloration of the kernels and, in extreme cases, an unpalatable "off-flavor". The moisture content of almonds at harvest ranges between 5% and 15% of fresh weight. To improve stability and ensure the safety of the nuts, they should be dried as soon after harvest as possible to 5–8% moisture or a water activity of 0.50–0.65. A pickup machine drives (Fig. 1.8) over the rows of nuts (Fig. 1.9) formed by

Fig. 1.8. Picking up the almonds after drying on the ground. Photo courtesy of Dr. Carlos H. Crisosto.

Fig. 1.9. Almond rows ready to be picked. Photo courtesy of Dr. Carlos H. Crisosto.

the sweeping operation. Nuts are picked up and conveyed into an attached trailer, while fine soil and leaves are blown out. The latest equipment has augers in the trailer that level the load and a conveyer that is activated by pressure on the rear bumper. The trailer conveyer unloads the trailer into a bank-out wagon when it applies pressure to the bumper by approaching from the rear. The bank-out wagon (Fig. 1.10), a specialized, low-profile hydraulic dump truck equipped with augers to level the load, ferries nuts out of the orchard directly to an on-farm huller, to a stockpile (Fig. 1.11) for later hulling, or to a set of double trailers (Fig. 1.12) at the roadside for a trip to a remote huller. Meanwhile, the pickup machine moves forward continuously, picking up the row of nuts, creating a harvest operation that

Fig. 1.10. Bank-out wagon. Photo courtesy of Dr. Carlos H. Crisosto.

Fig. 1.11. Almond stockpiles. Photo courtesy of Dr. Carlos H. Crisosto.

Fig. 1.12. Loading the double trailer. Photo courtesy of Dr. Carlos H. Crisosto.

is nearly non-stop. This series of harvest operations is repeated separately for each cultivar grown in the orchard. In California, almond orchards contain at least two cultivars for cross-pollination and many orchards have three or four cultivars.

Extra drying

Most of the final drying of almonds occurs naturally while they are on the orchard floor. However, when rainy conditions prevail during harvesting, heated-air drying can be used to complete their dehydration. Wet almonds must be dried before hulling either in a batch dryer with a maximum temperature of 48.9–54.4°C or in continuous-flow dryers with a maximum temperature of 80.2°C. Many batch dryers are specially designed, five ton capacity wagons with a perforated floor that allows heated air to be distributed underneath the nuts. The continuous-flow dryers are either horizontal belt dryers or crossflow grain dryers. Dried almonds hull more easily than those with wet hulls, allowing hulling equipment to operate at maximum capacity. Recently, an off-ground harvesting practice approach is being validated in California. Proposed innovative low-dust drying methods for almond harvesting including new shakers and/or an umbrella-type harvester which can be compatible for off-ground harvesting scenarios are being tested. A cost–benefit analysis using data from various sources is being used to evaluate the economic feasibility of switching to off-ground harvesting.

Hulling–Stockpile–Drying

The bank-out wagon transports almonds out of the orchard directly to an on-farm huller, to a stockpile for later hulling, or to a set of double trailers at the roadside for a trip to a remote huller. Because limited huller capacity leads to stockpiling of nuts, they will likely be covered and subjected to fumigation to limit insect damage. Nuts must be dried to <10% moisture prior to stockpiling. The in-hull nuts will be temporarily stored in stockpiles where nuts are covered with plastic tarps and fumigated with aluminum phosphide to control worms and preserve quality (Thompson et al., 1996). This practice also protects some of the crop from early fall rain and allows the huller to work out of the stockpile if rain delays other harvest operations. Once removed from the orchard, a mechanical huller, either owned by the producer or operated on a cooperative or custom basis, removes the hulls.

Physiological Disorders

Concealed damage (CD)

CD is defined as a dark brown discoloration of the kernel interior (nutmeat) that appears only after moderate to high heat treatment during processing of kernels with high moisture. Damaged kernels turn dark inside and a bitter taste can be perceived that can result in immediate consumer rejection. Only in very extreme cases, are kernel internal color and flavor altered prior to roasting. The economic damage is most apparent in cut kernels after they have been blanched, dried and roasted, so it makes it very difficult to predict and study (Kader, 1996; Rogel-Castillo et al., 2015, 2016, 2017). CD in kernels may be triggered prior to processing during harvest and/or after harvest when kernels are in windrows or stockpiles and exposed to rain or warm and moist environments. However, almond kernels with CD have no visible defects on the interior or exterior surface of the kernel prior to roasting. Even after roasting on CD-damaged kernels, there are no visible signs of CD on the surface of whole roasted kernels. Visual inspection and manual sorting are difficult, time-consuming, subjective, labor-intensive and cannot be used to identify nuts with CD prior to heat treatments (Rogel-Castillo et al., 2015). Based on current knowledge, CD is related to moisture-induced hydrolysis of sugars, elevated levels of volatiles related to lipid peroxidation, and amino acid degradation that can serve as reactants in the Maillard browning reaction occurring during processing (Severini et al., 2000; Rogel-Castillo et al., 2015; Lin et al., 2016). Further support for this hypothesis is the fact that aldehydes, such as 4,5-epoxy-2-alkenals, that have been detected in almonds, react with lysine amino groups to produce N-substituted hydroxyalkylpyrroles that polymerize spontaneously to form melanoidin-like pigments. Thus, almond kernels with CD have less near-infrared (NIR) absorbance in the region related to oil, protein, and carbohydrates as compounds have been hydrolyzed. The feasibility of using NIR spectroscopy partial least squares discriminant analysis (PLS-DA) models between 1125 nm and 2153 nm for the non-destructive and fast detection of CD in raw almonds was demonstrated (Rogel-Castillo *et al.*, 2016). Among these models that identify almonds with CD with 90.8–91.8% accuracy, the PLS-DA model based on the second-derivative spectra and using four wavelength ranges (i.e. 1408–1462, 1692–1740, 1902–1959, and 2064–2104 nm) related to the degradation of lipids, carbohydrates, and proteins gave the lowest percentage rate of false negatives and would be the best choice for further method development (Rogel-Castillo *et al.*, 2016).

Recent well-controlled and detailed studies demonstrated postharvest moisture exposure resulting in a kernel moisture content ≥8% is a key factor in the development of CD in almonds and that increases in temperature will accelerate this process (Rogel-Castillo *et al.*, 2015, 2017). Normally, these moisture content and temperature conditions that trigger CD occur when harvested nuts are stockpiled and fumigated to control NOW prior to hulling and shelling. Temperatures in covered stockpiles that are open to the sun can reach 60°C. If nuts have not been dried to <10% moisture in the orchard or have been wetted by late-season rains, the combination of elevated moisture and temperature lead to this problem. However, high temperature alone on kernels with low moisture does not cause the problem. Wetting of freshly harvested almonds followed by heating can cause the problem and forced-air drying of rain-wetted kernels can prevent it. Because research revealed that stockpile management can prevent aflatoxins and reduce CD, several handling recommendations were established. The first step to manage CD and *Aspergillus* growth is to ensure harvest moisture levels are met: the recommended harvest moisture levels for in-shell kernels are below 6%, in-hull almonds (total nut) are below 9% and hulls (alone) are below 12%. In general, almonds with the highest moisture levels are on the north side of the lower interior canopy near the tree trunk, where almond moisture tends to be about 2% higher than other areas in the tree canopy. Also, within the windrow, moisture tends to accumulate in the bottom layer of almonds. Thus, a practical guideline is do not stockpile if either the hull moisture content exceeds 12% or the kernel moisture content exceeds 6%. If rain is in the forecast, do not shake as almonds dry faster on the tree than on the ground. If rain comes after you have already shaken and nuts are still wet, blow or move them out of the tree row, but do not windrow. If almonds have been windrowed, pick them up and condition them through a drop chute for further drying. In all cases remove leaves and debris to help nuts dry faster. Achieving optimum moisture levels before stockpiling is essential to limit mold growth and CD. If moisture levels are higher than recommended to stockpile, move almonds to a dry area or dry by machine. Select an area for the stockpile that can be raised or sloped to encourage moisture to drain away from the stockpile to further limit mold growth. Stockpiles should be oriented with the long axis north–south and build with an even and flat top to minimize areas where condensation can build up on the underside of the tarp. When the long axis of stockpiles is oriented east–west, condensation and mold are usually worse on the north side of the pile. When tarps are used to cover stockpiles, tarps can increase humidity inside the stockpile triggering mold growth and CD. Use a white-on-black tarp that limits temperature fluctuations thus reducing condensation. A clear tarp should be used only on almonds that are well dried below the moisture threshold as it increases temperature fluctuations and condensation. White tarps without black had intermediate performance compared with the other two tarp types. As water activity is the best predictor of mold activity and food safety but is difficult to measure, relative humidity (RH) measurements are used to calculate water activity indirectly for whole almond, hulls and kernels. In this way by controlling the RH or water content in the stockpile, we can manage water activity that is critical for food safety.

To improve RH and water activity management in almond stockpiles, a relationship between RH and almond water activity was studied and a practical chart created (Almond Board of California, 2014). This table allows water activity of a stockpile and tissue water content to be calculated based on the RH measurement. An RH of greater than 65% within the pile is the maximum for almond storage. In the chart, green-shaded areas indicate moisture contents that are suitable for stockpiling. Yellow areas are borderline, and red areas indicate moisture contents that are too wet for stockpiling. For example, if the RH is too high in the stockpile, open the tarps in the daytime to allow moisture to escape, and close at night when RH tends to increase. Also look for condensation inside the stockpile due to large changes in temperature outside the stockpile. The use of this information assures that stockpiles are managed to exclude contaminants such as aflatoxins, harmful bacteria and pests, and to prevent mold-inducing moisture, and will help maintain almond quality and safety.

Pathological Problems

Most infections by pathogens are initiated in the orchard and because clean-up following harvest is not absolute, potential problems are transferred to the postharvest environment. In-shell product is relatively protected unless the shell has been broken or penetrated by insects. The most serious pathogens are fungi, such as *Aspergillus flavus* and *Aspergillus parasiticus*, which can produce aflatoxins that are both toxic and carcinogenic. Damaged kernels must be discarded prior to storage; controlled, low temperature and relative humidity conditions must be maintained. A variety of fungi are found on almonds, primarily species of *Aspergillus*, but also *Alternaria*,

Rhizopus, Cladosporium, and *Penicillium,* and these can be minimized by pasteurization, maintaining low moisture and water activity (Phillips *et al.,* 1979). Because of the relationship between insect damage and pathogen presence, sorting to eliminate whole kernels with insect damage reduces the number of kernels with excess concentrations of aflatoxins, considered to be concentrations greater than $1 \, ng \, g^{-1}$.

Aflatoxin

Aflatoxins are chemicals produced by certain molds that are of health concern because of their potential to cause cancer. Because of this health risk, maximum allowable levels of contamination in foods have been set in the market to protect consumers. Contamination can start in the orchard, as the fungus producing aflatoxins lives in the soil and can transfer to damaged mummy nuts that remain on the tree after harvest (Schatzki, 1996). From there, spores can be moved from nut to nut by NOW and/or other insects. Upon arrival to export markets, almonds tested with aflatoxins exceeding the allowable levels will be rejected and must be reconditioned at a high cost. A voluntary sampling program to prevent problems upon arrival is being established. Several pasteurization methods were developed to provide a safe, pathogen-free product without altering the flavor and nutritional characteristics of almonds. These sanitation technologies include commercially available physical processes such as blanching, oil roasting, steaming and treatment with propylene oxide (PPO) gas (Perren and Escher, 2013). PPO is a compound approved by the US Food and Drug Administration to pasteurize food products such as nuts, cocoa powder, and spices. Currently, almond pasteurization is required by law in the USA, Canada, and Mexico. In addition to postharvest processes to prevent outbreaks, a food quality and safety program is in place. It includes education programs for growers and handlers on Good Agricultural Practices (GAPs) in our orchards and Good Manufacturing Practices (GMPs) as a complete approach to provide consumers with the safest possible almonds.

Optimum Storage Conditions

Federal regulations define a safe moisture level for nuts as a water activity (aw) of <0.7 at 25°C to retard microbial growth. Because of the high content of unsaturated fatty acids found in almonds, like most nuts, they are susceptible to oxidation and to important quality losses if stored improperly or for too long. In general, almonds maintain quality and safety throughout storage better than some other nuts, due to their low moisture and high antioxidant content, thus their postharvest shelf life is primarily defined by changes in sensory attributes. The maintenance of appropriate texture (crispness and chewiness), absence of rancidity and odors related to oxidation levels provides an

Fig. 1.13. Shelled almonds storage in bins. Photo courtesy of Dr. Carlos H. Crisosto.

acceptable consumer quality. Furthermore, because of their high lipid content, almonds, especially shelled, should not be stored with commodities that have strong odors (onion, garlic, etc.) as they will absorb their odors during storage, resulting in less acceptability to consumers. Almonds are marketed in a variety of forms such as in-shell, shelled kernels, and peeled seeds; whole or nut pieces; and raw and roasted nuts, that influence product stability due to oxidation or rancidity development (Shahidi and John, 2013). Light or dark almond roasting is also relevant to stability; this common thermal process is used to create specific flavor notes, darken color, and add a more desirable crispy texture. Typically, the moisture content and water activity (aw) are reduced while the exposure to heat during roasting tends to directly increase rates of lipid oxidation. The process also produces Maillard reaction products with antioxidant properties that slow subsequent lipid oxidation in stored almonds. Current industry practices include storing raw in-shell almonds in silos, bins (Fig. 1.13), or other unlined cardboard (Fig. 1.14) cartons (UC) or bulk containers. During commercial storage roasted or raw almond kernels are stored either in polypropylene bags (PPB) or high barrier bags (HBB) across different combinations of temperatures and environmental RH. In-shell almonds can be stored longer than shelled almonds at the same temperature, indicating that the shell acts as a protective layer. Almonds can be packed using UC that provide no protection against transmission of water vapor or oxygen, or in plastic liners to protect from water losses. Thus, in some cases, almond processors are using PPB and/or HBB. Packaging of raw

Fig. 1.14. Shelled almonds storage in cardboard boxes. Photo courtesy of Dr. Carlos H. Crisosto.

shelled almonds in PPB, rather than UC, demonstrated substantial improvements in stability as measured by peroxidase values, moisture content, and water activity. Storing roasted almonds in HBB rather than PPB improved kernel quality stability as measured by peroxidase values, moisture content, and water activity. Thus, the use of HBB is a superior packaging choice, followed by PPB, with UC being associated with the greatest rates of degradation. The choice of packaging will be dictated by economics and the marketing conditions for which the almonds are handled. The low water content and high unsaturated fat contents of the kernel make it relatively stable metabolically and able to tolerate low temperatures, thus, the primary objective of any storage condition is to maintain the low water activity using temperature and RH. Current ideal recommendations are to store almonds under cool, dry conditions (<10°C) and <65% RH. The RH (safe levels) should be maintained to keep ~2.8–7.0% moisture in the kernels with a water activity (aw) of 0.2–0.8 at 20°C. At higher water activities, protection against fungal and human pathogens is lost. At very low water activity levels, sensory attributes such as texture, flavor, and color can be affected (Guiné *et al.*, 2015; Pleasance *et al.*, 2018; Wu *et al.*, 2019). In general, in-shell almonds can be stored for up to 20 months at 0°C, 16 months at 10°C, and 8 months at 20°C without a significant decrease in quality. Shelled almonds can be stored for about half the time of almonds in the shell (about 6 months). In all cases, avoid exposure to direct sunlight and protect from insects and pests, bad odors, and from oxygen, either through nitrogen flushing, bags, plastic liners, and/or vacuum packaging (Kader, 1996; Gama *et al.*, 2018).

Special Storage Treatments (Controlled Atmosphere and Others)

Shelled almonds (kernels) and roasted products are less stable than in-shell almonds and flavor is maintained well and longer under a low-oxygen and elevated-carbon dioxide atmosphere (Guadagni *et al.*, 1978; Kader, 1996; Raisi *et al.*, 2015). For example, flavor was maintained for 12 months at 18°C and 27.5°C in insect-controlling atmospheres of less than 1.0% oxygen and 9–9.5% carbon dioxide. Reduced oxygen in the storage atmosphere improves oil stability (Guadagni *et al.*, 1978; Kader, 1996). The stability difference between in-shell nuts and shelled kernels was eliminated in 0.5% oxygen. Storage containers, sealed plastic bags, and vacuum packaging are all good options that should be evaluated by the processor (Kader, 1996; Lin *et al.*, 2012; Martín *et al.*, 2016; Parrish *et al.*, 2019).

Quarantine Issues

The most serious almond postharvest insect problem is the NOW (*A. transitella*). This moth of the Pyralidae family is native to the south-western USA and Mexico and is a commercial pest of walnuts, figs, pistachios, and almonds. The insect lays its eggs in newly split nuts just before harvest, and the resulting larvae can cause substantial losses. Fumigation with methyl bromide, which has a limited future, or phosphine is used to control the insect. In-home control of insects can be attained by freezing temperatures of –5 to –10°C for a few days. Irradiation can also be used (300 Gy). The most useful non-chemical insect control approach appears to be a <1% O_2 and 9–9.5% CO_2 controlled atmosphere regime.

Almond Snack Products

Almonds are sold whole, as slices or flakes, slivers or halves, diced or chopped, meal or flour, almond milk, paste or butter, almond oil, and green almond. All of them except oil and green almond are available natural or blanched and are a popular snack or ingredient. These are used for confectionery, energy bars, trail mix or granola bars, cereals, yogurts, bakery products and inputs for processing. Whole almonds (natural or blanched) are widely used in their natural form or roasted and/or flavored, and even covered with chocolate. Slices or flakes (natural or blanched) are typically used as a topping for salads, an ingredient for cereal, a coating for savory dishes, a garnishing for baked goods, and for desserts. Slivers or halves (natural or blanched) work well as roasted or flavored snacks, are ideal for stir fries and grain dishes, as an ingredient for baked goods and cereal, to create texture for confectionery, and as a topping for prepared foods and salads. Diced or chopped (natural or blanched) forms were developed as toppings for dairy items and baked goods, and they work well for stuffing, as coatings for ice cream bars, filling for bakery products and confectionery, and as a crust for meats and seafood. Meal and flour (natural or blanched) are typically used as sauce thickeners, making almond butter or marzipan, as an ingredient and filling for confectionery, or as a flavor enhancer in baked goods and a coating for fried foods. Paste and butter (natural or blanched) provide an alternative to other spreads used as a filling for chocolate, cereal bars, confectionery and baked goods. Recently, almond milk sold in stores has become popular to be consumed directly in cereal or coffee, or blended into smoothies or lattes. Almond oil has been blended into vinaigrette salad dressing and has been used as non-food in cosmetics and moisturizers. Some consumers like eating green almonds by cutting the almond hull along the seam and extracting the immature, fresh, herbaceous-tasting kernel which is used as part of a mixed salad or eaten plain with a bit of sea salt.

Cull Utilization

Almond fruit consists of four tissues: (i) the kernel (embryo); (ii) a thin leathery seedcoat layer (pellicle); (iii) the middle shell (endocarp); and (iv) the outer green shell cover (exocarp, hull). Most of the nutritional importance of almond fruit is based on its kernel while hulls are rich in fiber and phytochemicals. Presently, hulls are used as livestock feed while shells are burned as fuel or used as livestock bedding. In the past decade, different phenolic compounds were characterized and identified in almond shell and hull tissues as potential almond by-products. These almond by-products are abundant in polyphenols that are important micronutrients in the human diet, and evidence for their role in the prevention of degenerative diseases such as cancer and cardiovascular diseases is emerging. The health effects of polyphenols depend on the amount consumed and on their bioavailability. Various phenolic compounds present in almond and its by-products, their antioxidant properties and potential use as natural dietary antioxidants, as well as their other beneficial compounds and applications are being reviewed. For example, 'Nonpareil' hulls contain 5-O-caffeoylquinic acid (chlorogenic acid), 4-O-caffeoylquinic acid (cryptochlorogenic acid), and 3-O-caffeoylquinic acid (neochlorogenic acid) in the ratio 79.5:14.8:5.7 and sterols. The chlorogenic acid concentration of almond hulls was 42.52 ± 4.50 mg 100 g^{1} of fresh weight (moisture content = 11.39%) having higher antioxidant activity than α-tocopherol and potentially similar antioxidant activity to chlorogenic acid and morin (2-(2,4-dihydroxyphenyl)−3,5,7-trihydroxy-4H-1-benzopyran-4-one) standards (at the same concentrations). The data indicate that almond hulls are a potential source of these dietary antioxidants. Presently, almond hulls are mainly used as a by-product for livestock feed. Several studies are being carried out to expand hull utilization, increase uses in dairy cow diets, expand to other livestock, and produce protein-rich feed additives for insect larvae and yeast growth. Almond by-products or biomass (shells, hulls, pruning, leaves, pellicles, and inedible kernel disposition) are being utilized but their potential values have not been fully reached. The almond by-products have been widely studied for bioenergy production using thermochemical and biochemical conversion technologies (Aktas *et al.*, 2015). With a large amount of almond shells produced annually and accumulated at huller and sheller facilities, California can have a steady supply for many units of thermochemical or biochemical conversion technologies across the Central Valley to operate year-round. In addition to this, the potential values of coproducts, such as biochar for soil amendment and bioliquids for biopesticidal applications or biochemicals, are being studied. Almond woody biomass as composite fillers and absorbents and torrefied shells as low-grade plastic enhancers are being explored for scale-up or commercial applications.

Research Needs

Research needs include:

- evaluation of new harvesting systems such as using an umbrella-type approach to avoid ground drying, reduce dust pollution, and assure consumer food safety;
- placing additional research emphasis on commercial conversion technologies that transform almond residues into energy and other valuable products;
- continuing the search for phytochemicals and antioxidant activities of almond hulls; and
- further study of the potential value of by-products such as biochar for soil amendment and bioliquids for biopesticide applications or biochemical uses.

References

Aktas, T., Thy, P., Williams, R.B., McCaffrey, Z., Khatami, R. *et al.* (2015) Characterization of almond processing residues from the central Valley of California for thermal conversion. *Fuel Processing Technology* 140, 132–147. DOI: 10.1016/j.fuproc.2015.08.030.

Almond Board of California (2014) Stockpile Management Best Practices Guide. Available at: https://www.almonds.com/sites/default/files/2020-04/grower_

stockpile_management_best_practices_from_abc_2014.pdf (accessed 20 August 2020).

Cheely, A.N., Pegg, R.B., Kerr, W.L., Swanson, R.B., Huang, G. *et al.* (2018) Modeling sensory and instrumental texture changes of dry-roasted almonds under different storage conditions. *LWT – Food Science and Technology* 91, 498–504. DOI: 10.1016/j.lwt.2018.01.069.

Connell, J.H., Sibbett, G.S., Labavitch, J.M. and Freeman, M.W. (1996) Preparing for harvest. In: Micke, W.C. (ed.) *Almond Production Manual.* Division of Agriculture and Natural Resources, University of California, Davis, California, pp. 254–259.

Connell, J.H., Labavitch, J.M., Sibbett, G.S., Reil, W.O., Barnett, W.H. *et al.* (1989) Early harvest of almonds to circumvent late infestation by navel orange worm. *Journal of the American Society for Horticultural Science* 114, 595–599.

Franklin, L.M. and Mitchell, A.E. (2019) Review of the sensory and chemical characteristics of almond (*Prunus dulcis*) flavor. *Journal of Agricultural and Food Chemistry* 67(10), 2743–2753. DOI: 10.1021/acs.jafc.8b06606.

Franklin, L.M., Chapman, D.M., King, E.S., Mau, M., Huang, G. *et al.* (2017) Chemical and sensory characterization of oxidative changes in roasted almonds undergoing accelerated shelf life. *Journal of Agricultural and Food Chemistry* 65(12), 2549–2563. DOI: 10.1021/acs.jafc.6b05357.

Franklin, L.M., King, E.S., Chapman, D., Byrnes, N., Huang, G. *et al.* (2018) Flavor and acceptance of roasted California almonds during accelerated storage. *Journal of Agricultural and Food Chemistry* 66(5), 1222–1232. DOI: 10.1021/acs.jafc.7b05295.

Gama, T., Wallace, H.M., Trueman, S.J. and Hosseini-Bai, S. (2018) Quality and shelf life of tree nuts: a review. *Scientia Horticulturae* 242, 116–126. DOI: 10.1016/j.scienta.2018.07.036.

Guadagni, D.G., Soderstrom, E.L. and Storey, C.L. (1978) Effect of controlled atmosphere on flavor stability of almonds. *Journal of Food Science* 43(4), 1077–1080. DOI: 10.1111/j.1365-2621.1978.tb15237.x.

Guiné, R.P.F., Almeida, C.F.F., Correia, P.M.R. and Mendes, M. (2015) Modelling the influence of origin, packing and storage on water activity, colour and texture of almonds, hazelnuts and walnuts using artificial neural networks. *Food and Bioprocess Technology* 8(5), 1113–1125. DOI: 10.1007/s11947-015-1474-3.

Kader, A.A. (1996) In-plant storage. In: Micke, W.C. (ed.) *Almond Production Manual.* Division of Agriculture and Natural Resources, University of California, Davis, California, pp. 274–277.

Kester, D.E. and Ross, N.W. (1996) History. In: Micke, W.C. (ed.) *Almond Production Manual.* Division of Agriculture and Natural Resources, University of California, Davis, California, pp. 1–2.

King, A.D. and Schade, J.E. (1986) Influence of almond harvest, processing and storage on fungal population and flora. *Journal of Food Science* 51(1), 202–205. DOI: 10.1111/j.1365-2621.1986.tb10870.x.

King, A.D., Halbrook, W.U., Fuller, G. and Whitehand, L.C. (1983) Almond nutmeat moisture and water activity and its influence on fungal flora and seed composition. *Journal of Food Science* 48, 615–617.

Lin, J.-T., Liu, S.-C., Hu, C.-C., Shyu, Y.-S., Hsu, C.-Y. *et al.* (2016) Effects of roasting temperature and duration on fatty acid composition, phenolic composition, Maillard reaction degree and antioxidant attribute of almond (*Prunus dulcis*) kernel. *Food Chemistry* 190, 520–528. DOI: 10.1016/j.foodchem.2015.06.004.

Lin, X., Wu, J., Zhu, R., Chen, P., Huang, G. *et al.* (2012) California almond shelf life: lipid deterioration during storage. *Journal of Food Science* 77(6), C583–C593. DOI: 10.1111/j.1750-3841.2012.02706.x.

Martín, M.P., Nepote, V. and Grosso, N.R. (2016) Chemical, sensory, and microbiological stability of stored raw peanuts packaged in polypropylene ventilated bags and high barrier plastic bags. *LWT - Food Science and Technology* 68, 174–182. DOI: 10.1016/j.lwt.2015.12.031.

Parrish, D.R., Pegg, R.B., Kerr, W.L., Swanson, R.B., Huang, G. *et al.* (2019) Chemical changes in almonds throughout storage: modeling the effects of common industry practices. *International Journal of Food Science & Technology* 54(6), 2190–2198. DOI: 10.1111/ijfs.14127.

Perren, R. and Escher, F.E. (2013) Impact of roasting on nut quality. In: Harris, L.J. (ed.) *Improving the Safety and Quality of Nuts.* Woodhead Publishing, Cambridge, UK, pp. 173–197.

Phillips, D.J., Mackey, B., Ellis, W.R. and Hansen, T.N. (1979) Occurrence and interaction of *Aspergillus flavus* with other fungi on almonds. *Phytopathology* 69(8), 829–831. DOI: 10.1094/Phyto-69-829.

Pleasance, E.A., Kerr, W.L., Pegg, R.B., Swanson, R.B., Cheely, A.N. *et al.* (2018) Effects of storage conditions on consumer and chemical assessments of raw 'Nonpareil' almonds over a two-year period. *Journal of Food Science* 83(3), 822–830. DOI: 10.1111/1750-3841.14055.

Raisi, M., Ghorbani, M., Sadeghi Mahoonak, A., Kashaninejad, M. and Hosseini, H. (2015) Effect of storage atmosphere and temperature on the oxidative stability of almond kernels during long term storage. *Journal of Stored Products Research* 62, 16–21. DOI: 10.1016/j.jspr.2015.03.004.

Reil, W., Labavitch, J.M. and Holmberg, D. (1996) Harvesting. In: Micke, W.C. (ed.) *Almond Production Manual.* Division of Agriculture and Natural Resources, University of California, Davis, California, pp. 260–264.

Rogel-Castillo, C., Zuskov, D., Chan, B.L., Lee, J., Huang, G. *et al.* (2015) Effect of temperature and moisture on the development of concealed damage in raw almonds (*Prunus dulcis*). *Journal of Agricultural and Food Chemistry* 63(37), 8234–8240. DOI: 10.1021/acs.jafc.5b03121.

Rogel-Castillo, C., Boulton, R., Opastpongkarn, A., Huang, G. and Mitchell, A.E. (2016) Use of near-infrared spectroscopy and chemometrics for the nondestructive identification of concealed damage in raw almonds (*Prunus dulcis*). *Journal of Agricultural and Food Chemistry* 64(29), 5958–5962. DOI: 10.1021/acs.jafc.6b01828.

Rogel-Castillo, C., Luo, K., Huang, G. and Mitchell, A.E. (2017) Effect of drying moisture exposed almonds on the development of the quality defect concealed damage. *Journal of Agricultural and Food Chemistry* 65(40), 8948–8956. DOI: 10.1021/acs.jafc.7b03680.

Sathe, S.K., Seeram, N.P., Kshirsagar, H.H., Heber, D. and Lapsley, K.A. (2008) Fatty acid composition of California grown almonds. *Journal of Food Science* 73(9), C607–C614. DOI: 10.1111/j.1750-3841.2008.00936.x.

Schatzki, T.F. (1996) Distribution of aflatoxin in almonds. *Journal of Agricultural and Food Chemistry* 44(11), 3595–3597. DOI: 10.1021/jf960120j.

Severini, C., Gomes, T., De Pilli, T., Romani, S. and Massini, R. (2000) Autoxidation of packed almonds as affected by Maillard reaction volatile compounds derived from roasting. *Journal of Agricultural and Food Chemistry* 48(10), 4635–4640. DOI: 10.1021/jf0000575.

Shahidi, F. and John, J.A. (2013) Oxidative rancidity in nuts. In: Harris, L.J. (ed.) *Improving the Safety and Quality of Nuts.* Woodhead Publishing, Cambridge, UK, pp. 198–229.

Thompson, J.F., Rumsey, T.R. and Connell, J.H. (1996) Drying, hulling and shelling. In: Micke, W.C. (ed.) *Almond Production Manual.* Division of Agriculture and Natural Resources, University of California, Davis, California, pp. 268–273.

USDA Agricultural Research Service (2019) FoodData Central. United States Department of Agriculture, Washington, DC. Available at: https://fdc. nal.usda.gov/fdc-app.html#/food-details/170567/nutrients (accessed 5 November 2019).

USDA Marketing Service (2019) Almonds in the Shell Grades and Standards. United States Department of Agriculture, Washington, DC. Available at: https://www.ams.usda.gov/grades-standards/almonds-shell-grades-and-standards (accessed 5 November 2019).

Wu, J., Lin, X., Lin, S., Chen, P., Huang, G. *et al.* (2019) California almond shelf life: changes in moisture content and textural quality during storage. *Transactions of the ASABE* 62(3), 661–671. DOI: 10.13031/trans.12709.

Fresh Fig

2

Carlos H. Crisosto[1]*, Margarita López Corrales[2], Giancarlo Colelli[3], and Manuel Joaquín Serradilla[4]

[1]*University of California, Davis, California, USA*
[2]*Centro de Investigación, Finca la Orden-Valdesequera, Badajoz, Spain*
[3]*University of Foggia, Foggia, Italy*
[4]*Instituto Tecnológico Agroalimentario de Extremadura, Badajoz, Spain*

Scientific Name, Origin and Current Areas of Production

The genus *Ficus*, of the *Moraceae* family, comprises ~700 species located mainly in the tropics, with six subgenera characterized by a reproductive system (Berg, 2003; Flaishman *et al.*, 2008; Khadari *et al.*, 2008). The cultivated fig tree, *Ficus carica* L. (2n = 26), belongs to the *Eusyce* section of the *Moraceae* family and is the only *Ficus* species cultivated for its fruit as a source of human food. This species was considered native to the Middle East, although recent studies show *F. carica* L. evolved from *F. carica* var. *rupestris*, which spread throughout the Mediterranean Basin before being domesticated at several simultaneous selection points in that area, including the Iberian Peninsula and the Balearic Islands. The fig tree was cultivated by all Mediterranean and Fertile Crescent civilizations for millennia. It may be the first fruit species cultivated by humans: there is archaeobotanical evidence that places its anthropic use 14,000 years ago in the Jordan Valley (Kislev *et al.*, 2006). Figs arrived in the New World in 1520, brought by the Spaniards, and in 1769 were introduced to California from Mexico by the Franciscan missionaries (Crisosto *et al.*, 2010, 2011). Today, figs for dried and fresh consumption are grown in many parts of the world with moderate climates, especially Mediterranean zones with mild winters and hot dry summers. Good conditions for fig growth are low relative humidity (RH) (<25%), intense sunlight with high summer temperatures (~32–37°C), and moderate winter temperatures.

---

*Corresponding author: chcrisosto@ucdavis.edu

Figs are harvested worldwide on 315,530 ha, with production of >1,153 million t in 2017. Turkey is the main producing country, with 23% of world production, followed by Egypt, Morocco, Algeria, and Iran, which together have 38% of global production (FAOSTAT, 2019). In Europe, Spain is the main producing country, with 35% of European production and 3.5% of world production. Turkey is the world's leading exporter of dried and fresh figs, although the marketing of these two products is carried out independently. Iran, Syria, and the USA are large exporters of dried figs while Austria, Italy, and Holland control exports of fresh figs. Spain, like Turkey, exports a considerable volume of both dried and fresh figs. The main destination for fresh fig exports is Europe and the dried fig market is more globalized. Among the more important commercial flows, the one between Turkey and some northern European countries such as Germany, Austria, and France stand out, reaching 80,000 t year[1]. These countries, together with Italy, control distribution, since they are not only important consumers, but also sell a large part of their imports from Turkey, Brazil, and South Africa to the rest of the continent. They are the primary source of supply to markets in Russia, Ukraine, Norway, and Finland, moving a volume of >20,000 t and generating ~60 million euros. US imports of dried whole figs come mainly from Turkey, Greece, and Mexico, while its imports of fig paste come mainly from Spain, Portugal, and Turkey. Canada, Japan, and Hong Kong are the three main markets for US dried fig exports. The USA ranks number seven in fig production, with 3.5% of the world's total. In 2017, the USA produced 40,212.5 t on 2590 ha, with yields three times the global average. Figs are produced commercially in 14 states; however, fig production is mainly concentrated in California (98%).

Fruit Morphology and Types

The fig fruit is borne from a complex inflorescence called a syconium (a "false fruit"), which encloses hundreds of fruits (true fruit; Fig. 2.1). The true fruits are drupelets derived from flowers inside the syconium. The tiny flowers and even the initial prosyconium are so small that figs were considered to bear fruit without ever forming flowers (Crisosto *et al.*, 2017). The syconium is connected to the exterior through a small aperture, depending on variety, of little scales termed the fig ostiole or eye (Fig. 2.2). Figs are divided into four groups depending on their sex and pollination. These four fruit type groups are: (i) Caprifig (*F. carica* var. *sylvestris* Shinn.); (ii) common fig (*F. carica* var. *hortensis* Shinn.); (iii) Smyrna type (*F. carica* var. *smyrnica* Shinn.); and (iv) San Pedro type (*F. carica* var. *intermedia* Shinn.). Common fig, Smyrna, and San Pedro are the only edible types, because they contain only long-styled female

Fig. 2.1. The fig fruit is a syconium (false fruit) which encloses hundreds of fruits (true fruit). Photo courtesy of Dr. Carlos H. Crisosto.

Fig. 2.2. The syconium is exposed to the environment through a small aperture (ostiole). Photo courtesy of Mr. Mike Poe, University of California Agriculture and Natural Resources (UC ANR), Davis, California, USA.

Fig. 2.3. Fig pollination of the Smyrna type fig is carried out by a wasp. Photo courtesy of Dr. Carlos H. Crisosto.

flowers, which produce high numbers of succulent fruitlets and function as females. The Caprifig type contains staminate and short-styled female flowers and acts as male to pollinate the female figs. Some Caprifigs are considered edible, with a more succulent fruitlet than typical Caprifigs. The common fig type has completely parthenocarpic or persistent flowers and produces fruit without pollination, the Smyrna type (for example 'Calimyrna') has completely non-parthenocarpic flowers and needs fertile pollination to produce fruit, and the San Pedro type needs pollination for the main crop, while the breba crop is parthenocarpic. Botanists use the term "persistent" rather than parthenocarpic when referring to the fig, since it is a false fruit with multiple fruits inside. Common and San Pedro fig types usually have two crops a year. Smyrna types usually have one crop, while Caprifig types bear three crops a year. As the name "common fig" suggests, a high proportion of fig cultivars (78% of those listed by Condit and Swingle, 1947) are "common". Fig pollination, which is also called "caprification", is carried out by the fig wasp (*Blastophaga psenes* L; Fig. 2.3), which has coevolved with the fig. In California, the main common fig varieties are 'Kadota', 'Black Mission', 'Adriatic', 'Conadria', and 'Brown Turkey'; the main Smyrna variety is 'Calimyrna'; the only two commercial Caprifig varieties are 'Roeding 3' and 'Stanford'; and San Pedro fig types are not commercially important (Flaishman *et al.*, 2008). In Spain, the majority of cultivated fig trees are persistent or parthenocarpic ones, whether they are annuals or biennials, and, to a lesser extent because there are fewer varieties, the San Pedro type. 'Colar', 'Coll Dame Negro' and 'Cuello Dama Blanco' are the primary varieties for fresh consumption and 'Calabacita', for dry consumption (M. Lopez-Corrales and M.J. Serradilla, Spain, 2019, personal communication). Smyrna type are not used because of their greater labor requirements. In Italy most of the commercial varieties are also parthenocarpic with 'Petrelli' and 'Brogiotto' being the main crops for fresh consumption (the former mainly as a breba crop), while 'Dottato' is the main crop for drying. In the countries of North Africa and the Middle East, Smyrna type varieties predominate and, to a lesser extent, San Pedro types.

Maturity, Harvest Index and Quality

Fruit growth, development and maturation

Fig syconium growth occurs at three different stages, described by a double sigmoid curve. The first stage (period I), during the first 6 weeks of growth, is characterized by rapid diameter increase and slightly less rapid increases in moisture, fresh and dry weight, and sugar content. The second growth stage (period II), during the successive 4 weeks, is characterized by reduced diameter growth, moisture content, and fresh and dry weight accumulation rates. During this stage, sugar content is relatively unaltered and ethylene production begins. The third growth stage (period III), during the remaining 4 weeks prior to maturation, is characterized by a marked increase in diameter, rates of fresh and dry weight increase, and moisture and sugar content. During this final growth stage, the fruit accumulates over 90% of its mature sugar content and >70% of its dry weight. At the beginning of phase III, the main epicuticular wax accumulation takes place in the form of regularly layered platelets. As ripening and senescence proceeds, wax quantity decreases and platelet structure changes, always maintaining a regular pattern. Finally, an early ethylene production peak is also observed at the beginning of phase III, coinciding with the maximum peak of ethylene at the commercially mature stage. Fig shape is usually spherical, cucurbiform, obovoid, turbinate, urceolate or pyriform (Fig. 2.4). Its commercial size may range from 25 mm to 70 mm long. Fig color varies depending on the variety; typical colors include yellow, green-yellow, yellow-green, copper, red, purple, and black, with or without stripes. The skin is tender and thin and the wall of the syconium is fleshy and either yellow-white, brown-yellow, amber, light pink, strawberry red, or purple (Flaishman et al., 2008). Figs are considered commercially mature when the figs are almost full color and the flesh gives slightly when touched (Fig. 2.5), while tree-ripe fruit (advanced maturity) is riper and softer (Fig. 2.6) than commercial maturity, but not overripe. In general, tree-ripe fig cultivars are visually more fully colored than those harvested at commercial maturity. A few tree-ripe fruits that were well exposed in the canopy can have leakage from the ostiole (Crisosto et al., 2011).

Fig. 2.4. Fruit size, color, and shape variation across cultivars. Photo courtesy of Dr. Carlos H. Crisosto.

Maturity stage and quality attributes

Maturity stage at harvest affects fig total soluble solids (TSS), titratable acidity (TA), consumer acceptance, firmness and loss of firmness during storage, ethylene production, respiration rate, shelf life, ostiole diameter, and shriveling (Table 2.1).

In addition, the amount of latex (Fig. 2.7) in the fig decreases as fresh figs ripen, increasing consumer acceptance (Crisosto *et al.*, 2010; Sortino *et al.*, 2017). In a study of ten different fig varieties, figs harvested at tree-ripe maturity had greater TSS and TSS:TA ratios, lower TA, greater consumer acceptance, and lower initial firmness than figs harvested at commercial maturity. Tree-ripe figs typically had lower ethylene production and respiration rates than figs harvested at commercial maturity. This decrease in ethylene production from commercial maturity to tree-ripe maturity occurs because the fig climacteric peak occurs prior to commercial maturity. Figs harvested at commercial maturity had a longer shelf life, but lower consumer acceptance, than those harvested at tree-ripe maturity, remaining sound and free of decay and off-colors for a longer shelf life after a week of cold storage (Crisosto *et al.*, 2010). Tree-ripe figs also had larger ostiole diameters and were more affected by shriveling and decay than figs harvested at commercial maturity. Maturity stage at harvest did not influence fig weight, size, antioxidant capacity, skin color and thickness, split ostioles, or blemishes.

Fig. 2.5. Commercial maturity. Photo courtesy of Dr. Carlos H. Crisosto.

Maturity stage influences flavor: for instance, overripe figs can develop undesirable flavors due to fermentation products (Ingrassia *et al.*, 2017). Fig firmness has a direct effect on the percentage of sound fruit, while the percentage of fruit with juice in the ostiole increases with the percentage of decayed fruit. "In store" consumer tests demonstrate that the degree of liking by consumers and percentage of consumer acceptance is directly related to TSS and the TSS:TA ratio, which have become two important parameters used to select fresh fig varieties (Table 2.2).

Other quality indices include absence of defects (such as bird-peck, sunburn, scab, skin break, and stem shrivel), insects, and fungal rot. Cultivar selection, fruit maturity and postharvest technology during marketing should be evaluated to protect flavor and increase consumer consumption. A consumer test evaluation on four fresh fig cultivars at two fruit maturity stages determined that the degree of liking was affected by cultivar and maturity stage at harvest, but there was no significant interaction

Fig. 2.6. Advanced maturity. Photo courtesy of Dr. Carlos H. Crisosto.

between cultivar and maturity stage. Figs harvested at tree-ripe maturity have significantly greater acceptance than figs harvested at commercial maturity, because they have developed their optimum organoleptic characteristics. The average of all tested cultivars for the tree-ripe figs reached 86% consumer acceptance, while commercial mature figs had only 66% acceptance. As many consumers are still unaware of fresh figs, educational promotion should be pursued due to the large potential for the fresh fig market.

Sensory descriptors for 12 fig cultivars (fresh and dried markets) growing in six locations-sources were evaluated to develop a sensory terminology (descriptors) and/or flavor code system. Figs at two maturities were evaluated by a trained panel using descriptive sensory analysis and at the same time instrumental measurements were taken at harvest and during sensory analysis. Each fresh fig cultivar had a characteristic appearance (Fig. 2.8) and flavor sensory profile regardless of the source. The primary flavor

Table 2.1. Interaction between cultivar and maturity stage on quality attributes of four fresh fig cultivars harvested from a commercial orchard in Madera County, California, in 2006. From Crisosto et al., 2010.

Cultivar	Maturity stage	Weight (g)	FTA (N)	Color[a]			TSS (%)	TA (% citric acid)	TSS:TA
				L^*	Chroma	Hue°			
'Mission'	Commercial	37.5	1.24	34.13	10.30	165.97	15.9	0.44	38.1
	Tree-ripe	35.6	0.85	30.86	4.03	271.21	19.1	0.38	51.0
'Brown Turkey'	Commercial	44.3	1.07	37.04	18.75	30.12	15.9	0.28	56.9
	Tree-ripe	52.2	0.65	31.23	11.71	71.24	18.0	0.29	62.4
'Calimyrna'	Commercial	52.7	2.29	69.95	54.31	112.00	15.7	0.62	25.8
	Tree-ripe	55.6	1.19	76.28	61.35	101.33	18.9	0.42	46.5
'Kadota'	Commercial	49.5	1.55	68.86	50.65	112.75	18.6	0.65	28.5
	Tree-ripe	54.9	1.04	74.86	53.12	105.54	19.3	0.22	86.1
P value		<0.0001	<0.0001	<0.0001	<0.0001	<0.0001	0.0009	<0.0001	<0.0001
$LSD_{0.05}$		4.9	0.26	2.47	3.18	61.81	1.8	0.12	14.07

FTA: fruit texture analyzer; LSD: least significant difference; TA: titratable acidity; TSS: total soluble solids.
[a]L^*, chroma and hue° are color parameters: lightness/darkness, chroma, and hue angle, respectively.

Fig. 2.7. Figs produce latex especially early during ripening. Photo courtesy of Dr. Carlos H. Crisosto.

Table 2.2. Consumer acceptance of four fresh fig cultivars harvested at commercial and tree-ripe maturity stages from a commercial orchard in Madera County, California, 2006. From Crisosto *et al.*, 2010.

	Degree of liking (1–9)[a][b]	Acceptance (%)	Neither like nor dislike (%)	Dislike (%)
Cultivar[b]				
'Mission'	6.3 b	72.0	8.5	19.5
'Brown Turkey'	5.7 a	63.5	11.0	25.5
'Calimyrna'	5.9 a	63.5	9.0	27.5
'Kadota'	6.8 b	81.5	5.0	13.5
Significance	0.0001			
Maturity stage[b]				
Commercial	5.3 a	65.8	12.0	33.2
Tree-ripe	7.0 b	85.5	4.7	9.8
Significance	0.0001			
Cultivar × maturity stage interaction	0.190			

[a]Degree of liking, on a scale of 1–9 where: 1 = dislike extremely; 2 = dislike very much; 3 = dislike moderately; 4 = dislike slightly; 5 = neither like nor dislike; 6 = like slightly; 7 = like moderately; 8 = like very much; and 9 = like extremely.
[b]Within cultivars or maturity stages, values labeled with different letters are significantly different ($P \leq 0.05$, Tukey's honest significant difference test).

Fig. 2.8. Trained panelist evaluation of fresh figs' visual attributes. Photo courtesy of Dr. Carlos H. Crisosto.

attributes used to describe the fig cultivars were "fruity", "melon", "stone fruit", "berry", "citrus", "honey", "green" and "cucumber" flavor. Maturity levels significantly affected the chemical composition and sensory profiles of the fig cultivars. Less mature figs had a higher compression force, a thicker outer skin, and higher ratings for "green" and "latex" flavors, firmness, graininess, bitterness, tingling, and seed adhesiveness. Meanwhile, more mature figs had higher TSS, and were perceptibly higher in "fruit" flavors, juiciness, stickiness, sliminess, and sweetness. The specific sensory terminology used for fig appearance and flavor profiles will assist with communication between marketers and consumers, which can increase fresh fig consumption (King *et al.*, 2012).

Composition and Health Benefits of Fig Fruits

F. carica L. are nutritious fruits rich in dietary fiber, potassium, calcium, and iron, with higher concentrations of these nutrients than other common fruits such as bananas, grapes, oranges, strawberries, or apples. Figs are sodium-free, fat-free, and like other fruits, cholesterol-free. Additionally, figs are a good source of vitamins (thiamine and riboflavin), amino acids (aspartic acid and glutamine), and antioxidants. Compounds with antioxidant properties, such as vitamin C, tocopherols, carotenoids, and polyphenols, can "modify the metabolic activation and detoxification/disposition of carcinogens", affect processes that alter the development of tumor

Table 2.3. Antioxidant capacity of four fresh fig cultivars (*n* = 3) harvested at commercial and tree-ripe maturity stages from a commercial orchard in Madera County, California, 2006. From Crisosto *et al.*, 2010.

	TEAC (µmol TE g^{-1} FW)[a]
Cultivar	
'Mission'	3.14 a
'Brown Turkey'	1.73 bc
'Calimyrna'	1.88 b
'Kadota'	1.44 c
Significance	0.0001
Maturity stage[a]	
Commercial	1.98 a
Tree-ripe	2.12 a
Significance	0.088
Cultivar × maturity stage significance	0.160

FW, fresh weight; TE, Trolox equivalents; TEAC, Trolox equivalent antioxidant capacity.
[a]Within cultivars and maturity stages, values labeled with different letters are significantly different (*P* ≤0.05, Tukey's honest significant difference test).

cells, and avoid neurochemical and behavioral changes related to aging (Wojdyło *et al.*, 2016). The common fig is an old food that has been a human staple since ancient times (Barolo *et al.*, 2014): it is mentioned in the Gospels, the sacred book of Christians. Figs have been used as traditional medicine to treat a wide variety of ailments, alone or in combination with drugs (Allegra *et al.*, 2017). Fruits and vegetables rich in phenolic compounds can decrease cardiovascular and cerebrovascular diseases and cancer death rates. Dried figs are rich in polyphenols, such as anthocyanins, proanthocyanidins, flavonols, and flavanones, containing higher concentrations than most fruit and beverages consumed today. Fig varieties with dark skins contain more polyphenols, accompanied by greater antioxidant activity, than fig varieties with lighter skins (Table 2.3).

Most of compounds with antioxidant activity, such as anthocyanins, are located in the fig skin rather than the flesh. Cyanidins (claimed to be the only anthocyanins in figs) are the main pigment compound in skin, with cyanidin-3-*O*-rutinoside the primary anthocyanin responsible for the characteristic skin color (Flaishman *et al.*, 2008).

Fruit Physiological Characteristics

Fresh figs are very perishable climacteric fruits, susceptible to mechanical damage and highly susceptible to weight loss, softening, and postharvest decay infections that significantly reduce their cold storage and shelf life. Fresh figs produce very low amounts of ethylene and carbon dioxide. Respiration measured as carbon dioxide (CO_2) evolution is dependent on cultivars and temperature.

Respiration and carbon dioxide production

Respiration was measured in popular fig cultivars, and carbon dioxide production varied from $3\,ml\ CO_2\ kg^{-1}\ h^{-1}$ at $0°C$, increased to $10\,ml\ CO_2\ kg^{-1}\ h^{-1}$ at $10°C$ and to $25\,ml\ CO_2\ kg^{-1}\ h^{-1}$ at $20°C$. Similar temperature effects were measured for ethylene (C_2H_4), low ethylene levels at $0°C$ ($0.6\,\mu l\ C_2H_4\ kg^{-1}\ h^{-1}$) and at $5°C$ ($1.2\,\mu l\ C_2H_4\ kg^{-1}\ h^{-1}$). As the temperature rises, C_2H_4 production increases to $2.0\,\mu l\ C_2H_4\ kg^{-1}\ h^{-1}$ at $10°C$ and $5.0\,\mu l\ C_2H_4\ kg^{-1}\ h^{-1}$ at $20°C$ (Crisosto et al., 2011).

Ethylene sensitivity

Responses to ethylene have their climacteric peak at the beginning of the third growth stage (period III), when figs are still hanging on the tree. Preharvest ethylene applications to figs have different effects depending on the stage of fruit development. Ethylene applied in period I inhibits fruit growth and triggers fruit abscission. Ethylene applied in period II stimulates fruit growth and eventually, abscission. In addition, such application also stimulates external color change, but fruits do not mature completely, have a mealy texture and lack sweetness and flavor. Ethylene applied late in period II or in period III stimulates growth and maturity. Ethylene has been used to regulate maturity uniformity since ancient times. Applications of olive oil on the ostiole 10 days after the drupelets turned red ensured the presence of ethylene, produced by degradation of the oil. Application of exogenous abscisic acid (ABA) enhances fig fruit ripening and maintains ripening uniformity (Lama et al., 2019). Therefore, both ethylene and ABA control ripening in fresh fig. Postharvest ethylene applications (continuous addition of 10 ppm) on 'Brown Turkey', 'Kadota' and 'Sierra' fig cultivars did not affect the quality of fruit stored at $0°C$, $5°C$ or $20°C$. Ethylene exposure only affected the percentage of purple skin color on immature 'Brown Turkey' figs stored at $20°C$, increasing purple skin color from 8.3% to 100% in 7 days.

Physiological Disorders and Chilling Temperature Sensitivity

Chilling injury or other physiological disorders have not been observed or reported in fig, except for sunburn. Ostiole-end splitting (Fig. 2.9) and fruit skin side cracking damage are important factors in losses because

Fig. 2.9. Ostiole-end splitting. Photo courtesy of Dr. Carlos H. Crisosto.

they affect decay development and the percentage of sound fruit during fresh fig postharvest handling and marketing. The type and degree of skin damage varied among cultivars. For 'Brown Turkey', 'Kadota' and 'Sierra', slight skin damage prior to cold storage increased decay and reduced postharvest life. In contrast, the postharvest life of 'Black Mission' fig was not significantly affected by a slight degree of skin damage prior to cold storage. Furthermore, postharvest decay incidence was associated with the degree of side cracking and ostiole-end splitting at harvest. Because fruit skin side cracking and ostiole-end splitting occur during fruit growth and development, prevention by regulated deficit irrigation (RDI) with 55% ETc (evapotranspiration of the crop) was studied for 2 years (Kong *et al.*, 2013). In both seasons, fruit quality attributes were not affected by RDI, except for 'Brown Turkey', where size decreased by 21% during one season. RDI significantly reduced fruit skin side cracking and ostiole-end splitting in 'Brown Turkey' and skin side cracking in 'Sierra', increasing marketable fruit by 50% in 'Brown Turkey' and 18% in 'Sierra' (Kong *et al.*, 2013).

Postharvest Handling

Harvest operations

Brebas are the first figs of the season, setting on wood from the previous year, and they typically mature in June in California's Central Valley. In south-western Spain and south-eastern Italy the main fig crop is produced

Fig. 2.10. Fresh fig trees are kept short to allow pickers to harvest from the ground without using ladders. Photo courtesy of Mr. Mike Poe, UC ANR, Davis, California, USA.

on the current season's wood, maturing fruit from August to September, or even later in a warm year. Fresh figs are harvested manually based on firmness and color change (Fig. 2.10). Figs at harvest should give a little to the touch, but still be firm. Dark-skinned varieties should be harvested before turning completely dark, while green-skinned varieties should be harvested when they are yellowish-white to light yellow. Fresh fig skin color and flesh firmness are related to their quality and postharvest life. Achievement of ripeness in the fig crop is sequential: the first fig fruits to ripen are those at the base of the new shoots and they ripen consecutively towards the distal end of the shoot. In fresh fig production, this sequential ripening makes multiple harvest dates necessary to harvest the fruit at its optimal time. Fresh figs are usually harvested daily, every other day, or weekly for 4–6 weeks, depending on cultivar. In contrast, the breba crop achieves ripeness over a shorter period.

Fresh fig harvesting must be done carefully, minimizing the physical damage, abrasions, and cracks that make fruit more susceptible to decay. To avoid damage, figs should be harvested early in the morning using harvesting scissors, detaching the fruit with a clean cut that retains the stem and avoids lesions. Using gloves (Fig. 2.11) while managing figs helps reduce fruit damage and bruising and protects the worker's skin from the latex (caustic milky exudate which causes skin irritation) released from the broken stem. Fresh figs must be harvested exclusively from the tree,

Fig. 2.11. Fresh figs are harvested using gloves and clippers. Photo courtesy of Dr. Carlos H. Crisosto.

never from the ground. In addition, containers used at harvest and during transportation require previous strict sanitation to reduce disease exposure. Canning figs are picked when fully colored, but still firm. Harvest is by hand, often from ladders, and the figs are placed in buckets or shallow flats. Fruits are usually removed from the branch in one motion, by grasping the fig in the hand while twisting and pulling. Fresh figs are highly perishable, so it is exceedingly important to transport (Fig. 2.12) the harvested product without delay to a packing area and then to the cold storage facility.

Packinghouse Practices

Fresh figs are usually packed at the side of the field after harvest (Fig. 2.13), to avoid having the fruit under pressure for a long time. Packing takes place in sheds to avoid excessive heat that would reduce shelf life. Figs are packed in containers that allow air circulation in one-layer plastic-pack trays with 15–70 fruit in a ~8.1-lb container (Fig. 2.14). In general, there are eight large figs or 12–16 small figs per pound. When clamshells are used, 12 8-oz clamshells are placed in the container (Fig. 2.15). Containers can be reusable plastic containers (RPC) or cardboard boxes of 12 in long × 8

Fig. 2.12. Fresh figs are transported frequently to the packing area. Photo courtesy of Dr. Carlos H. Crisosto.

in wide × 2.5 in high. The trays, containers or clamshells are then put into cardboard boxes and stacked on pallets for transport. After packing, fresh figs must be cooled as soon as possible to −0.5–0°C. It is very important to maintain the cold chain during packing, storage, and transportation to improve fig shelf life. Thus, the cold chain must be maintained throughout handling until fruit reaches the consumer.

Temperature Handling

In 'Kadota' and 'Brown Turkey' figs, weight loss (WL) correlated strongly with delayed hauling or the onset of cooling due to transpiration and respiration. After arrival at cold storage, figs that were cooled immediately (i.e. 0 h delay (0HD)) had lost only 0.5% of their fresh weight, while there was 5–6% WL in figs left 3 h (3HD) or 6 h (6HD) in the sun before storage at 0°C, depending on cultivar. For 13 days subsequent cold storage, figs handled under the 0HD treatment lost 2% and 4% of their weight, after 6 and 13 days, respectively. After 13 days cold storage, 6HD figs had 12% WL. After 6 days cold storage, 'Kadota' and 'Brown Turkey' figs with fast cooling (0HD and 3HD) were firmer than figs that began cooling after 6 h (6HD) of field delay (data not shown). The 6% and 12% weight losses

Fig. 2.13. Field pack area at the side of the field. Photo courtesy of Mr. Mike Poe, UC ANR, Davis, California, USA.

imply a direct loss of six or 12 boxes per 100 boxes without considering figs lost due to quality blemishes. In both cultivars tested, shorter delays before cooling produced a greater percentage of sound fruit during shelf life. The no delay (0HD) or 3HD treatments produced more sound fruit than 6HD fruit during cold storage. After 6 days of storage at 0°C, 0HD and 3HD had a greater percentage of sound fruit (80–100%) than 6HD (40%). 'Kadota' figs from the 6HD treatment had a greater percentage of fruit with "off color" and other blemishes that reduced the sound category. However, after 2 days at store display temperature (20°C), "off color" increased in the 3HD treatment and decay incidence increased in the 6HD treatment, but not in other treatments. These changes decreased figs in the sound category and as a consequence, sound figs decreased to 66% in the 3HD and to 40% in the 6DH treatment, while 0HD treated figs had no losses. After 4 days at 20°C, fig losses increased dramatically (~60–80%): the percentage of sound figs was 44% for 0HD, 26% for 3HD and only 18% for 6HD (Crisosto *et al.*, 2011).

Optimum Storage and Shipping Conditions

'Mission' figs stored at 5°C had almost double the respiration rate of fruits stored at 0°C, while their ethylene production was almost 50-fold greater. Similar results occurred in 'Calimyrna'. Fresh fig shelf life, or the period in which the fruit maintains its commercial quality at room temperature, is

Fig. 2.14. Fresh figs are packed in containers that allow air circulation. Photo courtesy of Dr. Carlos H. Crisosto.

extremely short, lasting 1–2 days at 20°C after storage at 4–6°C and 85% RH (Crisosto *et al.*, 2011). Delays of 3 h or 6 h in starting cooling significantly increased weight loss, reduced firmness, lowered the percentage of sound figs and reduced postharvest life during cold storage and warm display (shelf life) of the varieties 'Kadota' and 'Brown Turkey'. Expedited transportation from field to cold storage and fast, forced-air cooling to −1–0°C is highly recommended. The optimal storage conditions for fresh figs are temperatures of −1–0°C with a RH of 90–95%. Fresh figs will absorb odors produced by apples and green onions.

Main Fresh Fruit Pathological Problems

Fresh figs with postharvest damage are very susceptible to decay by *Alternaria* spp., *Rhizopus* spp., *Botrytis* spp., *Cladosporium* spp., and other organisms, which significantly reduces their cold storage and shelf life (Fig. 2.16). *Aureobasidium* spp. is the primary yeast on different fig varieties during cold storage. Decay is an ever-present hazard in the postharvest handling of fresh figs, as surface contamination by fungal spores creates field infections that can be assumed to be present. Decay reduction on fresh figs is a difficult task, due in part to the lack of resistant genotypes and

Fig. 2.15. Fresh figs are packed in clamshells. Photo courtesy of Dr. Carlos H. Crisosto.

registered preharvest and postharvest products. The high susceptibility of figs to decay, combined with their fast senescence, which accelerates fruit softening, makes fresh fig postharvest life extremely short. As fresh fig marketing and consumption spreads to places far from centers of production, we have assessed the benefits of current postharvest technologies such as temperature management (fast hauling and cooling) and safe sulfur dioxide application to maximize their postharvest life.

Special Storage Treatments

Controlled atmosphere (CA)

CA and modified atmosphere packaging (MAP) with reduced oxygen concentration and increased carbon dioxide (not more than 25%) have beneficial effects on fresh fig storage and prolong their shelf life (Colelli *et al.*, 1991; Colelli and Kader, 1993; Crisosto *et al.*, 2011; Guillén *et al.*, 2015; Bahar and Lichter, 2018). Specifically, storage in a CA containing 5–10% oxygen (O_2) and 15–20% carbon dioxide (CO_2) is beneficial to reduce decay, maintain firmness, and reduce respiration and ethylene production rates. Reduced respiration and ethylene production rates lengthen fresh fruit life. Extended storage in a CA can result in loss of characteristic flavor.

Fig. 2.16. Postharvest decay mainly due to *Alternaria* spp. and *Cladosporium* spp. Photo courtesy of Dr. Carlos H. Crisosto.

Good quality of fresh 'Mission' figs was maintained for up to 4 weeks at 0°C, 2.2°C, or 5°C in atmospheres enriched with 15% or 20% CO_2. The visible benefits of high CO_2 concentrations were reduced decay incidence and maintenance of a bright external appearance. Ethylene production was reduced and fruit softening (measured using a deformation tester) was slower in figs stored in high CO_2 than in those kept in air. Similar results were observed for 'Calimyrna' figs. On the other hand, the acetaldehyde concentration of CO_2-treated 'Mission' fig fruit increased during the first week, then decreased, while ethanol increased slightly during the first 3 weeks and moderately during the fourth week. This could potentially result in "off-flavor". Exogenous ethylene exposure during cold storage does not affect fruit quality. For instance, continuous exposure of 10 ppm ethylene to 'Brown Turkey', 'Kadota' and 'Sierra' fig cultivars did not affect fruit quality, internal maturity, percentage of sound fruit (commercial fruit), percentage of fruit with decay, or percentage of fruit with off-color when stored at 0°C, 5°C or 20°C. However, exogenous 1-methylcyclopropene (1-MCP) improved storage quality of 'Albacor' fig by reducing storage disorders, fruit softening, and microbial infection during storage up to 21 days (Villalobos *et al.*, 2018). Modified atmospheres can be created with biaxially oriented polypropylene or polyethylene films by packing the fruits individually, in trays or in small containers (Villalobos *et al.*, 2014).

Fig. 2.17. Fresh figs treated with 25 ppm h^{-1} sulfur dioxide (left) and untreated (right) after 2 weeks in cold storage and 2 days at warm temperature. Photo courtesy of Dr. Carlos H. Crisosto.

Microperforated films can be a useful tool to maintain fig quality traits. Such films extended the postharvest life of 'Cuello Dama Negro', 'Black Mission' and 'San Antonio' figs for up to 21 days of cold storage without receiving any chemical treatments, while 'Cuello Dama Blanco' or 'Kadota' figs have an optimal storage time of up to 14 days under these conditions (Villalobos *et al.*, 2016). A combination of active MAP (polypropylene punnets under gas concentrations of 20 kPa CO_2 and 20 kPa O_2) with $CaSO_4$ powder used as moisture absorber (3% p/p) was a highly effective treatment to delay fungal development on breba fruit of the Italian 'Domenico Tauro' cultivar, maintaining counts <2 log cfu g^{-1}, well below all other treatments, although the presence of white powder on the fruit skin decreased their visual acceptability (Villalobos *et al.*, 2017).

Sulfur dioxide gas application

A supplementary postharvest sulfur dioxide (SO_2) gas application treatment to control decay was developed (Fig. 2.17). A 25 ppm h^{-1} (CT, concentration × time product) SO_2 gas treatment applied prior to or during cooling effectively controlled fig decay within their limited postharvest life (Cantín *et al.*, 2011). Our data confirm that proper temperature management is critical to extend postharvest life and maintain quality during global distribution. SO$_2$ treatment is an enhancement and not a solution to extend fresh fig marketing.

Other storage treatments

Another approach has been to apply a short exposure of high temperature (30 min at 50°C) to the 'Petrelli' cultivar. This non-chemical treatment reduced fruit decay at room temperature from 35.1% to 8.1% for a few days (Piazzolla et al., 2018). This partial decay control treatment can be useful in subtropical fig production areas such as Turkey, Greece, Brazil, Spain, and Egypt, where in most cases cooling of the product is not possible and the cold chain during distribution is very poor. Several researchers are validating the use of coatings combined with thymol on decay control and evaluating the impact on sensory characteristics (Saki et al., 2019).

Utilization

Culinary, medicinal and other uses

Figs are a very nutritious fruit that can be consumed either fresh or dried. Depending on the variety, dried figs have different uses such as for table consumption or for processing as paste, canned, jam, chocolate fig bonbons or alcoholic beverage production. Historical works provide evidence of the sustained importance and appreciation of figs in the Mediterranean area. Fig fruits, leaves, and latex were used for traditional medicines. Pliny the Elder in his *Natural History* extolled "one hundred and eleven observations" on the fig. Among them, "This fruit invigorates the young, and improves the health of the aged and retards the formation of wrinkles." Fig fruits have antipyretic, purgative, and aphrodisiac properties and are useful in treating inflammation and paralysis. Fig fruits have also been used against respiratory diseases. Fig leaves have been used against anemia and as an anthelmintic. Fig latex (milky substance produced by the fig tree) has been used against warts, corns, and insect bites. Figs have modern industrial applications. Ficin, a proteolitic enzyme from the fig shoots and the immature fruits, can be used as a meat tenderizer, as a chill proofing agent in the beer industry, as a milk coagulant, and for removing the casing of sausages. Efforts are ongoing to develop possible new fig products through the combination of different technologies including reduction of water activity (aw) and high-pressure processing. Very promising results related to physical, chemical, and microbial attributes were obtained when fruits were dipped for 15 min in a 0.5% ascorbic acid solution, partially dehydrated in cold air (up to 0.87 of water activity), then packaged in fructose isotonic solution and processed at 600 MFa for 3 min. Under these conditions, fruits retained more fresh-like attributes and had a lower microbial population than with other treatments (blanching, dehydration, and MAP) and kept these characteristics for 28 days at 5°C (Allegra *et al.*, 2017).

Suitability of fig fruit as a fresh-cut product

Fresh figs performed well as fresh-cut fruit and could add flavor and diversity to cut fruit marketing. Fresh-cut figs (maroon-skinned 'Brown Turkey' and yellow-green-skinned 'Sierra' varieties) maintained quality after 12 days in air at 0°C and in CA at 5°C with 12% or 18% CO_2. At 0°C storage, intact and cut figs had similar postharvest life, but storing sanitized fresh-cut figs was the best option. At 5°C, the high-CO_2 atmospheres provided some benefit and cultivar response differed (Hong *et al.*, 2016). Although decay was minimal, methods to better sanitize the fruit surface could enhance shelf life of cut figs. Ripeness at harvest affected fig texture and composition, affecting sensory quality of the fresh-cut fruit. Therefore, accurate determination of a relatively narrow window of maturity will be critical to prepare fresh-cut figs with good eating quality and shelf life. Temperature control was much more important than CA for fresh-cut fig shelf life.

Cull utilization

Fig culls and leaves can be used as animal feed.

Retail Outlet Display Considerations

Because of their high perishability and short market life, fresh figs should be displayed dry and at low temperature. Although in some Mediterranean areas, fresh figs are displayed at room temperature (Fig. 2.18).

Special Research Needs

There is a great deal of interest in expanding fresh fig sales in the USA, Europe, and Asia. This will require significant advances in postharvest handling. The large number of consumers still unaware of figs, combined with positive consumer perception, suggests a large, untapped potential for the fig market, especially the fresh fig market. To realize that potential, it is necessary to develop cultivars better suited for fresh consumption, with better flavor at the current commercial maturity stage and/or that remain firm enough at the tree-ripe stage to tolerate harvest, postharvest handling and marketing. In addition, development and establishment of technologies for extending commercial life and reducing losses of fresh figs in production areas where refrigeration equipment capacity is lacking, and the cold chain is non-existent are critical to increase fig consumption in the

Fig. 2.18. Warm fresh fig store display. Photo courtesy of Dr. Carlos H. Crisosto.

Mediterranean area. Areas of scientific research that will require funding to improve fresh fig consumption include: (i) handling of fresh-cut fruit; (ii) breeding superior cultivars; (iii) handling technologies to extend market life; (iv) health benefits; and (v) information outreach and marketing to protect safety and quality.

References

Allegra, A., Alfeo, V., Gallotta, A. and Todaro, A. (2017) Nutraceutical content in 'Melanzana' and 'Dottato' fig fruit (*Ficus carica* L.). *Acta Horticulturae* 1173, 319–322.

Bahar, A. and Lichter, A. (2018) Effect of controlled atmosphere on the storage potential of Ottomanit fig fruit. *Scientia Horticulturae* 227, 196–201. DOI: 10.1016/j.scienta.2017.09.036.

Barolo, M.I., Ruiz Mostacero, N. and López, S.N. (2014) *Ficus carica* L. (Moraceae): an ancient source of food and health. *Food Chemistry* 164, 119–127. DOI: 10.1016/j.foodchem.2014.04.112.

Berg, C.C. (2003) *Flora Malesiana* precursor for the treatment of Moraceae 1: the main subdivision of *Ficus*: the subgenera. *Blumea - Biodiversity, Evolution and Biogeography of Plants* 48(1), 166–177. DOI: 10.3767/000651903X686132.

Cantín, C.M., Palou, L., Bremer, V., Michailides, T.J. and Crisosto, C.H. (2011) Evaluation of the use of sulfur dioxide to reduce postharvest losses on dark and green figs. *Postharvest Biology and Technology* 59(2), 150–158. DOI: 10.1016/j.postharvbio.2010.09.016.

Colelli, G. and Kader, A.A. (1993) CO_2-enriched atmospheres reduce postharvest decay and maintain good quality in highly perishable fruits. *Proceedings of COST 94 - 3rd Workshop on Controlled Atmosphere Storage*, Milan, Italy, 22-23 April 1993, pp. 137–148.

Colelli, G., Mitchell, F.G. and Kader, A.A. (1991) Extension of postharvest life of 'Mission' figs by CO_2-enriched atmospheres. *HortScience* 26(9), 1193–1195. DOI: 10.21273/HORTSCI.26.9.1193.

Condit, I.J. and Swingle, W.T. (1947) *The Fig*. Chronica Botanica Co., Waltham, Massachusetts.

Crisosto, C.H., Bremer, V., Ferguson, L. and Crisosto, G.M. (2010) Evaluating quality attributes of four fresh fig (*Ficus carica* L.) cultivars harvested at two maturity stages. *HortScience* 45(4), 707–710. DOI: 10.21273/HORTSCI.45.4.707.

Crisosto, C.H., Ferguson, L., Bremer, V., Stover, E. and Colelli, G. (2011) Fig (*Ficus carica* L.). In: Yahia, E.M. (ed.) *Postharvest Biology and Technology of Tropical and Subtropical Fruits, Vol 3: Cocona to Mango*. Woodhead Publishers, Cambridge, UK, pp. 134–158.

Crisosto, C.H., Ferguson, L., Preece, J.E., Michailides, T.J., Haug, M.T. *et al.* (2017) Developing the California fresh fig industry. *Acta Horticulturae* 1173(1173), 285–292. DOI: 10.17660/ActaHortic.2017.1173.49.

Food and Agricultural Organization of the United Nations (FAOSTAT) (2019) Digital Report – the State of Food and Agriculture. Available at: http://www.fao.org/statistics/en/ (accessed 9 January 2020).

Flaishman, M., Rodov, V. and Stover, E. (2008) The fig: botany, horticulture and breeding. *Horticultural Reviews* 34, 113–196.

Guillén, F., Castillo, S., Valero, D., Zapata, P.J., Martínez-Romero, D. *et al.* (2015) Use of modified atmosphere packaging improves antioxidant activity and bioactive compounds during postharvest storage of 'Collar' figs. *Acta Horticulturae* 1071(1071), 263–268. DOI: 10.17660/ActaHortic.2015.1071.32.

Hong, G., Crisosto, C. and Cantwell, M.I. (2016) Quality and physiology of two cultivars of fresh-cut figs in relation to ripeness, storage temperature and controlled atmosphere. *Acta Horticulturae* 1141(1141), 213–220. DOI: 10.17660/ActaHortic.2016.1141.25.

Ingrassia, M., Chironi, S., Allegra, A., Sortino, G., Caruso, T. *et al.* (2017) Consumer preferences for fig fruit (*Ficus carica* L.) quality attributes and postharvest storage at low temperature by in-store survey and focus group. *Acta Horticulturae* 1, 383–388.

Khadari, B., Roger, J.P., Ater, M., Achtak, H., Oukabli, A. *et al.* (2008) Moroccan fig presents specific genetic resources: a high potential of local selection. *Acta Horticulturae* 798(798), 33–37. DOI: 10.17660/ActaHortic.2008.798.3.

King, E.S., Hopfer, H., Haug, M.T., Orsi, J.D., Heymann, H. *et al.* (2012) Describing the appearance and flavor profiles of fresh fig (*Ficus carica* L.) cultivars. *Journal of Food Science* 77(12), S419–S429. DOI: 10.1111/j.1750-3841.2012.02994.x.

Kislev, M.E., Hartmann, A. and Bar-Yosef, O. (2006) Early domesticated fig in the Jordan Valley. *Science* 312(5778), 1372–1374. DOI: 10.1126/science.1125910.

Kong, M., Lampinen, B., Shackel, K. and Crisosto, C.H. (2013) Fruit skin side cracking and ostiole-end splitting shorten postharvest life in fresh figs (*Ficus carica* L.), but are reduced by deficit irrigation. *Postharvest Biology and Technology* 85, 154–161. DOI: 10.1016/j.postharvbio.2013.06.004.

Lama, K., Modi, A., Peer, R., Izhaki, Y. and Flaishman, M.A. (2019) On-tree ABA application synchronizes fruit ripening and maintains keeping quality of figs (*Ficus carica* L.). *Scientia Horticulturae* 253, 405–411. DOI: 10.1016/j.scienta.2019.04.063.

Piazzolla, F., Amodio, M.L. and Colelli, G. (2018) Effects of thermal treatments on quality of 'Petrelli' figs during storage. *Acta Horticulturae* 1194, 879–888.

Saki, M., Valizadehkaji, B., Abbasifar, A. and Shahrjerdi, I. (2019) Effect of chitosan coating combined with thymol essential oil on physicochemical and qualitative properties of fresh fig (*Ficus carica* L.) fruit during cold storage. *Journal of Food Measurement and Characterization* 13(2), 1147–1158. DOI: 10.1007/s11694-019-00030-w.

Sortino, G., Gallotta, A., Barone, E. and Tinella, S. (2017) Postharvest quality and sensory attributes of *Ficus carica* L. *Acta Horticulturae* 1, 353–358.

Villalobos, M., Serradilla, M.J., Martín, A., Ruiz-Moyano, S., Pereira, C. *et al.* (2014) Use of equilibrium modified atmosphere packaging for preservation of 'San Antonio' and 'Banane' breba crops (*Ficus carica* L.). *Postharvest Biology and Technology* 98, 14–22. DOI: 10.1016/j.postharvbio.2014.07.001.

Villalobos, M., Serradilla, M.J., Martín, A., López-Corrales, M., Pereira, C. *et al.* (2016) Preservation of different fig cultivars (*Ficus carica* L.) under modified atmosphere packaging during cold storage. *Journal of the Science of Food and Agriculture* 96(6), 2103–2115. DOI: 10.1002/jsfa.7326.

Villalobos, M.C., Ansah, F., Amodio, M.L., Serradilla, M.J. and Colelli, G. (2017) Application of modified atmosphere packaging with moisture absorber to extend the shelf life of 'Domenico Tauro' breba fruit. *Acta Horticulturae* 1173(1173), 365–370. DOI: 10.17660/ActaHortic.2017.1173.63.

Villalobos, M.C., Serradilla, M.J., Ruiz-Moyano, S., Martin, A., López-Corrales, M. *et al.* (2018) Postharvest application of 1-methylcyclopropene (1-MCP) for preservation of 'Albacor' figs (*Ficus carica* L.). *Acta Horticulturae* 1194, 853–860.

Wojdyło, A., Nowicka, P., Carbonell-Barrachina, A.A. and Hernández, F. (2016) Phenolic compounds, antioxidant and antidiabetic activity of different cultivars of *Ficus carica* L. fruits. *Journal of Functional Foods* 25, 421–432. DOI: 10.1016/j.jff.2016.06.015.

Peach and Nectarine

3

Carlos H. Crisosto[1]*, Gemma Echeverría[2], and George A.
Manganaris[3]
1University of California, Davis, California, USA
2Institut de Recerca i Tecnologia Agroalimentaries (IRTA), Lleida, Spain
3Cyprus University of Technology, Lemesos, Cyprus

Scientific Name, Origin and Current Areas of Production

The peach (*Prunus persica* (L.) Batsch) is native to China. The nectarine
(*Prunus persica* (L.) Batsch var. *nectarina*) is a peach that lacks fuzz and is
caused by a specific natural gene mutation of peach. Chinese literature
dates peach cultivation in China to 1000 BC. Then, peach was transferred
from China to Persia (Iran) and at one time it was called a "Persian apple". Thereafter, it quickly spread to Europe. In the 16th century, the
Spaniards introduced peach cultivation to Mexico, while during the 18th
century Spanish missionaries introduced the peach to California, which
turned out to be one of the top ten most important peach (fresh and
canning) production areas. Currently, approximately 24,665 t of peaches and nectarines are produced worldwide, with China producing almost
60%, followed by the European Union (EU) (Spain, Italy, and Greece)
producing 15% and the USA *c.*3.3% (FAO, 2017). In general, traditional
nectarine fruit have higher acidity than peach fruit, a smaller size, and
a greater susceptibility to bacterial disease which limits their cultivation
(Crisosto and Costa, 2008). Over recent years, there has been a significant release of white- and yellow-fleshed peach cultivars with high total
soluble solids (TSS), a high aromatic profile and low acidity from an array of active breeding programs around the world (Crisosto *et al.*, 2001a;
Crisosto, 2002; Iglesias, 2019). In addition, peaches and nectarines with
different texture, flavor, flesh color, and shape became commercially
available in a continuous attempt to provide a wide selection with enhanced qualitative attributes and in certain cases with a low-softening
rate and adequate storage potential.

*Corresponding author: chcrisosto@ucdavis.edu

Maturity and Harvest Indexes

The maturity stage at which stone fruits, including peach and nectarine, are harvested greatly influences their ultimate flavor and market life. Harvest maturity controls the fruit's flavor components, physiological deterioration problems, storage potential, susceptibility to mechanical injury, resistance to moisture loss, susceptibility to rot organisms, market life, size and ripening capacity. Early-harvested peach fruit, when they are still immature, may fail to ripen properly (abnormal ripening) and the green skin background color may never fully disappear. Immature fruit typically soften slowly and irregularly, without reaching the desired melting texture and flavor of fully matured fruit (Crisosto and Valero, 2008). Since immature fruit lack a fully developed surface cuticle, they are more susceptible to water loss, being characterized by lower TSS and higher acidity than properly matured fruit. All the above-mentioned factors contribute to inadequate flavor development after ripening. Also, less mature harvested fruit are thereafter more susceptible to both abrasion and the development of chilling injury (CI) symptoms than properly matured fruit. Peaches harvested when they are over mature (onset of the senescence stage) are too soft to tolerate postharvest handling and shipping. Such fruits become highly susceptible to mechanical injury and fungal invasion, becoming overripe with poor eating quality, including "off-flavors" and a mealy texture (Ferrer *et al.*, 2005; Crisosto and Costa, 2008; Iglesias and Echeverría, 2009). Thus, the optimum maturity for harvest must be defined for each peach cultivar. The highest maturity at which a cultivar can be successfully harvested is influenced by postharvest handling, genotype, technology availability, temperature management procedures and marketing (Crisosto and Costa, 2008). Maturity selection and determination are more critical for distant markets than for local markets, but this does not necessarily mean lower maturity and/or quality (Crisosto and Valero, 2008).

Maturity indices based on skin background color and firmness when applied properly are useful to determine the right maturity. Supervision of the application of the maturity index during harvesting operations is critical to assure the potential market life of the lot. In California and other places, harvest date is determined by following skin background color changes from green to yellow in most cultivars. A color chip guide for background color is used to determine maturity of most cultivars except for white-fleshed and fully red-colored cultivars. Fruit skin background color is a useful, objective, non-destructive method for determination of fruit maturity, and is most easily employed and understood by field workers during harvesting operations (Crisosto *et al.*, 2004a; Ferrer *et al.*, 2005; Crisosto and Costa, 2008). In new cultivars that are characterized by intense fully red color where skin ground color is masked by early full red color development prior to maturation, fruit firmness or non-destructive tools are being used to determine how long fruit can be left on the tree before harvest (Crisosto and Valero, 2008; Drogoudi *et al.*, 2016).

Maximum maturity is defined as the minimum flesh firmness at which fruits can be handled without bruising damage (Crisosto *et al.*, 2001b, 2004a; Crisosto and Costa, 2008). Application of the maturity index such as skin background color and/or flesh firmness is carried out by the harvesting crew leader. Prior to harvest at the beginning of the block-cultivar, the crew leader meets with pickers to discuss and show them the kind of fruit that they should be harvesting. Proper description of the maturity index and easily understood directions for recognizing maturity should be given to the workers (Crisosto and Costa, 2008). By selecting a few fruits of varying maturity and demonstrating what maturity level is acceptable and unacceptable, many mistakes are avoided. It is recommended to leave these fruit samples with the crew leader as a reference throughout the day. When a maturity index based on fruit firmness is used on highly red-flushed-colored cultivars, the instructions to the harvesters will also imply minimum size and location of the fruit in the tree canopy (Crisosto *et al.*, 2001b; Crisosto and Costa, 2008). The value of a good and expert crew leader cannot be overemphasized. The crew leader's role is critical throughout the harvesting process for continuously monitoring the fruit being picked and the fruit remaining on the tree to determine if the correct balance is achieved. Orchard managers should involve the crew leaders in all stages of the decision-making process when determining optimum harvest maturity. Doing so will give the crew leader greater understanding and experience in the process, a key issue to optimize final consumer satisfaction. More importantly, it will solidify in the crew leader's mind the importance of their role in harvesting fruit at the proper maturity stage.

Quality and Consumer Preferences

"Fruit quality" is a concept that encompass sensorial properties (appearance, texture, flavor, taste, and aroma), nutritional value, and mechanical properties (Crisosto, 2002; Crisosto and Costa, 2008). Altogether, such attributes give the fruit a degree of excellence and an economic value (Delgado *et al.*, 2013). Everyone in the peach production and marketing chain from the grower to the consumer looks for fruit with no or few visual defects. However, in each step of this chain, the term "quality" takes on different meanings and the economic relevance of the various quality traits is largely variable. For example, the grower is interested in high yield, in fruit with large size and enhanced disease resistance, ideally combined with a reduced number of harvests that subsequently reduce labor costs.

The definition of "quality" for packers, shippers, distributors, and wholesalers is mainly based on flesh firmness that is considered a good index to predict fruit potential storage and market life. Peaches ripen and deteriorate quickly at ambient temperature and cold storage is required to slow down deterioration, especially for some cultivars and/or transportation to long-distance markets (Crisosto, 2002). In some production areas such as California, Chile, Italy, Spain, and South Africa, the development of CI symptoms such as lack of flavor, "off-flavor", mealy textured flesh and flesh browning limits the storage life and the postharvest quality of flavorful cultivars (Crisosto and Labavitch, 2002; Crisosto and Valero, 2008; Manganaris *et al.*, 2017). For retailers, red color, size, and firmness have historically represented the main components of fruit quality, as they need fruit that are attractive to the consumer, resistant to handling, and have a long shelf life. The grower, identifying fruit quality almost exclusively with the fruit size and color, does not consider that these are only the first characters perceived by the consumer and they orient only for his/her very first choice (Delgado *et al.*, 2013). As soon as consumers realize that the fruit, even with good size and attractive color, is tasteless, with low sugar content, poor aroma and is rapidly perishable, they redirect their interest towards other types of fruit. In general, peach fruit quality and consumption have declined, mainly because of premature harvesting, CI, improper handling, and lack of ripening prior to consumption, resulting in consumer dissatisfaction. In order to increase peach consumption, attention and application of the "consumer quality" concept (defined as a combination of peach traits that will satisfy consumers, mainly based on flavor and free of sensory defects) should be used as a main priority by growers and handlers (Crisosto, 2002; Crisosto and Tonutti, 2015). Consequently, it is imperative for the growers and other individuals in the delivery chain to direct their attention to fruit quality from the consumer's perspective ("consumer quality") in order to regain their confidence. In addition, nowadays there is an increasing understanding that "consumer quality" also includes nutritional properties (e.g. vitamins, minerals, dietary fiber), whether the fruit is easy to eat, and contains antioxidants and other health-promoting benefits (Byrne *et al.*, 2009) and these are becoming important factors in consumer preferences (Crisosto and Tonutti, 2015). Since the consumer quality of peaches does not improve after harvest, it is important to understand the role of orchard factors in consumer acceptance and postharvest life (Crisosto and Costa, 2008).

A generic descriptive analysis of the basic flavor sensory attributes, aroma, taste, and texture characteristics, were carried out by an experienced trained panel for two seasons. Twenty-three peach and 26 nectarine cultivars collected from four orchard sources were consistently classified into four predominant sensory flavor attributes groups: (i) balanced; (ii) tart; (iii) sweet; and (iv) peach flavor intensity (Crisosto *et al.*, 2006). Simultaneously, the "in-store" acceptability of the same cultivars was evaluated by USA consumers (Crisosto and Crisosto, 2005). Furthermore, cultivars within these groups were highly accepted or rejected by the consumers according to their ethnic preferences (Crisosto *et al.*, 2006). Based on the creation of a "flavor code" in cultivars that were classified into distinct sensory groups, we developed a minimum quality index within each "flavor code" group rather than proposing a generic minimum quality index based on the TSS. Thus, this "flavor code" cultivar classification helps to match ethnic preferences and enhance current promotion and marketing programs (Crisosto *et al.*, 2006; Crisosto and Tonutti, 2015).

In order to predict consumer acceptance, searching for drivers of likings, the correlations between instrumental measurements at harvest (flesh firmness, TSS, and titratable acidity), sensory panel descriptors, and consumer liking hedonic responses ("in store" consumer tests) were investigated. Cultivars with medium acidity and/or flavor-aroma were liked "very much", and consumer willingness to pay a premium was correlated with overall liking, irrespective of the cultivar considered. In this large group of peach and nectarine cultivars, cluster analysis revealed three segments that were associated with ethnicity and consumer preferences. Sweetness was the main driver of liking for two consumer clusters; however, for the third cluster, the perception of fruit aromas described as grassy/green fruit and pit aromas ("off-flavor") was the main driver of liking. There was a high correlation between instrumental measurements and their sensory perception; however, the sensorial attributes were more positively highly correlated with overall liking and consumer acceptance than instrumental measurements alone (Delgado *et al.*, 2013). In sound fruits, TSS was the only instrumental measurement significantly related to consumer liking (Crisosto and Crisosto, 2005; Crisosto *et al.*, 2006; Delgado *et al.*, 2013).

In California, as an industry guideline, a minimum of 10% TSS for yellow-fleshed peaches and nectarines is recommended as a quality standard on fruit free of sensory defects and CI (Crisosto and Valero, 2008). Similarly, in France a minimum of 10% TSS for medium-low acidity peaches and 11% TSS for high-acidity peaches ($\geq$0.9%) is being used successfully as part of their quality standards. In Italy, a minimum of 10% TSS for early-season, 11% TSS for mid-season and 12% TSS for late-season cultivars was suggested for yellow-fleshed peaches. In the EU, $\geq$8% TSS and $\leq$6.5 kgf fruit firmness was established for early- to late-ripening peaches by the European Commission Regulation that are notably below the current recommendations of $\geq$11% TSS and fruit firmness <5 kgf (Iglesias and Echeverría, 2009). A multi-country effort (France, Germany, Italy, Poland, and Spain) using consumer tests to compare selected peach cultivars that were different in type, flavor, flesh texture, and skin-flesh color revealed that consumers liked especially sweet, low acid, juicy, and aromatic peaches that had TSS values of 13.6% and a titratable acidity of less than 0.6% (Bonany *et al.*, 2014). For instance, in Spain, the second largest world producer, peach consumption per capita was only 3.8 kg year^{-1} while for Greece it exceeds 20 kg year^{-1} (Iglesias, 2019). Over the last 15 years, the Spanish peach industry has renewed the old orchards introducing new and improved cultivars in terms of color, flavor, and type of fruit (the "Spanish varietal revolution"), like the flat peaches that are additionally character-ized by high TSS and low acidity. Cultivars have been selected based on their high color, sweet flavor, size, crispiness, and slow melting texture, re-sulting overall in an increased benefit for both the growers and the con-sumers (Lavilla *et al.*, 2002; Iglesias, 2019).

Fruit Composition and Health Benefits

Peaches are characteristically soft-fleshed and highly perishable fruit with high flavor and limited market life. A peach fruit is *c.*87% water and the rest is carbohydrates, organic acids, pigments, phenolics, vitamins, volatiles, anti-oxidants, and trace amounts of proteins and lipids (Proteggente *et al.*, 2002; Vicente *et al.*, 2009; Drogoudi *et al.*, 2016; USDA, 2018). An immature peach fruit contains very low amounts of starch grains. Such grains are rapidly con-verted into soluble sugars as the fruit matures on the tree. Consequently, after harvest, there is no significant increase in soluble sugars during storage and subsequent ripening. Soluble sugars contribute approximately 7–18% of the total weight and fiber contributes *c.*0.3% fresh weight (FW) of the total fruit. Sucrose, glucose, and fructose represent about 75% of peach fruit soluble sugars (Moriguchi *et al.*, 1990). Malic acid is the predominant organic acid in mature peach fruit followed by citric acid (Crisosto and Valero, 2008; USDA, 2018). These organic acids (0.4–1.2% FW) are important because it has been reported that the ratio of soluble solids to titratable acidity determines con-sumer perception in most ripe peach cultivars (Crisosto *et al.*, 2006). Some of these cultivars have a very low acidity that for some ethnic groups is not considered a desirable attribute. In most mid–high acidity peach cultivars, acidity decreases by *c.*30% during postharvest ripening.

Peach fruit have low protein content (0.5–0.8% FW) but these small-size proteins have an important function as enzymes catalyzing an array of chemical reactions responsible for compositional changes. Despite lipids constituting only 0.1–0.2% FW, they are important because they make up the surface wax that contributes to the cosmetic appearance of the fruit and cuticle that protects the fruit against water loss and pathogens. Lipids are also important constituents of cell membranes, affecting an array of fruit physiological processes (Brummell *et al.*, 2004a).

Minerals in fruits include base-forming elements (Ca, Mg, K, and Na) and acid-forming elements (P, Cl, and S). Calcium is associated with cell wall structure and possesses a critical role in fruit softening, while apoplas-tic calcium has been related to senescence. Postharvest changes in fruit mineral content are negligible. Volatile compounds in very low concentra-tions include esters, alcohols, aldehydes, ketones, and acids, and these are responsible for the characteristic fruit aroma. Lactones may be sensorially important in peach "off-flavor" development during cold storage, but more detailed studies are needed.

The peach fruit is considered to be a good source of antioxidants, containing ascorbic acid (vitamin C), carotenoids (provitamin A) and phenolic compounds; their content is highly variable based on the cultivar considered (Tomás-Barberán *et al.*, 2001; Gil *et al.*, 2002; Goristein *et al.*, 2002; Drogoudi *et al.*, 2016). The total ascorbic acid (vitamin C) content in peach ranged from 6 to 9 mg 100 g^{-1} FW in white-fleshed cultivars and from 4 to 13 mg 100 g^{-1} FW in yellow-fleshed cultivars. Total carotenoids concentration was in the range of 71–210 µg 100 g^{-1} FW for yellow-fleshed and 7–20 µg 100g^{-1} FW for white-fleshed peach cultivars (Tomás-Barberán *et al.*, 2001; Goristein *et al.*, 2002). Thus, there were about tenfold more carotenoids in yellow-fleshed than in white-fleshed peach cultivars. The main carotenoid detected was β-carotene (provitamin A), but also small and variable quantities of β-cryptoxanthin are present in some white-fleshed peach and nectarine cultivars (Gil *et al.*, 2002). The total phenolics concentration varied from 28 to 111 mg 100 g^{-1} FW for white-fleshed and from 21 to 61 mg 100 g^{-1} FW for yellow-fleshed Californian cultivars. Phenolics have a role in the visual color appearance of fruit, taste (astringency) and contribute to antioxidant activity, which is potentially beneficial to health (Gil *et al.*, 2002). The predominant hydroxycinnamic acid is chlorogenic acid, and catechin and epicatechin are the main procyanidins identified and their concentrations are higher in white-fleshed than in yellow-fleshed peaches. Cyanidin-3-glucoside is the predominant anthocyanin, which, along with other anthocyanins, is present mainly in the skin. Concentrations of antioxidant flavonols (including quercetin and kaempferol) are higher in yellow-fleshed than in white-fleshed peaches (Tomás-Barberán *et al.*, 2001). The antioxidant capacity per peach fruit serving (100 g of peel and flesh) varied widely according to the cultivar considered. In general, white-fleshed peaches were slightly higher in total antioxidant capacity than yellow-fleshed peaches (Byrne *et al.*, 2009; Vicente *et al.*, 2009). Total antioxidant capacity ranged from 13 mg to 107.3 mg of ascorbic acid equivalents when evaluated by the DPPH (2,2 -diphenyl-1-picrylhydrazyl) free radical assay and from 19 mg to 119.6 mg of ascorbic acid equivalents when evaluated by the FRAP (ferric reducing ability plasma) method (Gil *et al.*, 2002).

Peach extracts have been described as showing suppression activity in breast cancer cell lines while having little effect on normal cells (Noratto *et al.*, 2009, 2014; Vizzotto *et al.*, 2014). Chlorogenic acid and neochlorogenic acid in the peach phenolic fractions where identified as the bioactive compounds with potential chemopreventive and/or chemotherapeutic natural compound properties (Noratto *et al.*, 2009, Noratto *et al.*, 2014). In addition, studies with peach pericarp extracts were conducted against cisplatin-induced acute toxicity in mice (Lee *et al.*, 2008). Results showed a significant protection against the acute nephrotoxicity and hepatotoxicity which is likely caused by reducing cisplatin-induced oxidative stress in mice. Recent studies with peach, nectarine, and apricot extracts have shown *in vitro* binding of a mixture of bile acids (secreted in human bile at a duodenal physiological pH of ~6.3) (Kahlon and Smith, 2007). Using cholestyramine (bile-acid-binding, cholesterol-lowering drug) as a reference indicated that binding potential follows the order peach > apricots > nectarines (Kahlon and Smith, 2007).

Fruit Physiological Characteristics

The peach fruit is botanically classified as a drupe which is soft-fleshed and highly perishable with a limited market life potential. A drupe is a fleshy fruit with a thin, edible outer skin (epicarp) derived from the ovary, an edible flesh of varying thickness beneath the skin (fleshy mesocarp), and a hard, inner ovary wall that is highly lignified (endocarp) and is commonly referred to as the stone or pit, which encloses a seed. Peaches have thin skins and soft flesh. The skin, as a protective layer, is composed of cuticle, epidermis, and some hypodermal cell layers. The cuticle is a thin coating of wax and serves to reduce water loss and to protect the fruit against mechanical injury and attack by pathogens. The epidermis, consisting of heavy-walled cells, is responsible for most of the skin's mechanical strength. Surface hairs ("fuzz") of peach fruit are extensions of some epidermal cells that are easily broken. The flesh, which is the main edible portion of the fruit, consists mainly of storage parenchyma tissue composed of large, relatively thin-walled cells with high water content. Based on separation of stone from flesh, nectarine and peach varieties can be divided into two groups: (i) freestone where the stone does not adhere to the flesh; and (ii) clingstone (where the stone adheres firmly to the flesh). Based on final flesh texture, peaches and nectarines can be classified as melting or non-melting, while based on their rate of ripening they can be classified as fast-ripening, slow-ripening, and in between. From the commercial point of view, clingstone peaches are usually non-melting, adaptable to the canning process. However, in Europe and other places, consumers enjoying eating as fresh fruit both freestone and clingstone cultivars.

Upon the completion of pollination and fertilization of the egg, the flower ovary begins to enlarge into a developing fruit. This is "fruit set" and it marks the beginning of growth and development. Stone fruits have a double sigmoidal growth curve which includes three distinct stages of growth. Following fruit set, cell division continues for about 4 weeks, with cell enlargement beginning and proceeding rapidly (Stage 1). Slow growth occurs during which lignification of the endocarp (pit hardening) and growth of the endosperm and embryo inside the seed take place (Stage 2). Cell enlargement (expansion) resumes in the flesh (mesocarp) tissue. The fruit continues to increase in size until it reaches full maturity, after which growth slows markedly and finally stops (Stage 3). The duration of each stage of growth depends upon cultivar, climactic conditions, and some cultural practices (such as thinning or crop load per tree, soil moisture, girdling, and nutrition). Fruit density (specific gravity) declines during Stage 1, increases during Stage 2, then declines again during Stage 3 (final fruit swell). During the pit hardening, the seed constitutes *c*.25% of the fruit weight, and this value drops to 14% during final swell. From a postharvest standpoint, interest in Stage 3 is greatest, since maturation, ripening, and senescence occur during this stage. Maturation is the time between final growth and the beginning of ripening. Maturity is the end point of maturation. An immature fruit may ripen off-tree, but it will be of poor quality. A mature fruit will attain good quality when ripened off the tree. Ripening involves changes that transform the mature fruit into one ready to eat. Changes associated with ripening include loss of green color and development of yellow, red, and other colors, typical of the cultivar. As a fruit ripens, it softens, its starch is converted to sugars, its acidity declines, and it produces certain volatile compounds that give it a characteristic aroma (Crisosto and Valero, 2008). Physical and chemical changes continue after "optimum" ripeness is reached (from a flavor quality standpoint), including further softening and loss of desirable flavor.

Rates of respiration and ethylene production

Increased respiration and ethylene production rates are among the physiological changes associated with ripening. Respiration measured as carbon dioxide (CO_2) evolution depends on the cultivar and postharvest temperature applied. In California-bred cultivars, respiration, measured at their peak, varied from 4–6 mg CO_2 kg^{-1} h^{-1} at 0°C, increasing to 16–24 mg CO_2 kg^{-1} h^{-1} at 10°C, and to 64–110 mg CO_2 kg^{-1} h^{-1} at 20°C. The range within each temperature indicates the commercial cultivar variability. Similar variation was measured for ethylene (C_2H_4) production in Californian commercial fruit. Low ethylene levels were recorded at 0°C (0.01–5 µl C_2H_4 kg^{-1} h^{-1}) and at 5°C (0.02–10 µl C_2H_4 kg^{-1} h^{-1}). As temperature rises, C_2H_4 production increases: to 0.05–50 µl C_2H_4 kg^{-1} h^{-1} at 10°C and 0.10–160 µl C_2H_4 kg^{-1} h^{-1} at 20°C (Crisosto and Valero, 2008). However, ethylene levels are not directly connected with the texture and softening pattern of each cultivar (Manganaris *et al.*, 2008).

Sensitivity to ethylene

As peaches are harvested mature but firm after climacteric respiration rise occurred, they do not respond and benefit from ethylene exposure. At the same time, they are not sensitive to ethylene exposure. Immature fruit with low consumer acceptance potential can be softened by exposing to ethylene and move throughout the marketing channels. These immature peaches will trigger consumer dissatisfaction that will affect total peach and nectarine consumption.

Fig. 3.1. Peaches are carefully selected and harvested by hand. Photo courtesy of Mr. Mike Poe, University of California Agriculture and Natural Resources (UC ANR), Davis, California, USA.

Harvesting Handling

Peaches and nectarines, as climacteric fruits, are harvested from April to October in California when they reach at least a minimum degree of maturity but firm to ensure consumer quality. Thus, they are not completely ripe ("ready-to-eat") and the beginning of the ripening process must occur prior to consumption to satisfy consumers (Crisosto and Valero, 2008). However, when consumers eat even high quality but unripe fruit, they will not be satisfied, and the consumers will wait a long time before buying again. Peaches and nectarines are hand-picked (Fig. 3.1) into bags, totes or buckets (Fig. 3.2). The fruit are dumped into bins that are on the top of trailers between rows in the orchard. Plastic bin liners and padded bin covers have been shown to reduce transport injury in some sensitive conditions. Totes are placed directly inside the bins and baskets are placed on modified trailers. Fruit picked at advanced maturity stages, as well as white-fleshed peaches or nectarines, are generally picked and placed into baskets or totes. Depending on the cultivar a worker can usually harvest one and a half to three full-size bins of fruit per day. Early-season cultivars are usually picked every 2–3 days, and by mid- to late-season, the interval can stretch to as much as 7 days between harvests. In general, early-ripening cultivars are harvested twice while mid- and late-ripening cultivars are harvested three to six times according to cultivar, season, and prices. Tree heights are

Fig. 3.2. General view of high-quality peaches picked in buckets. Photo courtesy of Dr. Carlos H. Crisosto.

commonly 3.7–4.7 m, and workers require ladders to reach the uppermost fruits. The recent establishment of pedestrian orchards that include different training and pruning and the use of size-control rootstocks are reducing the use of ladders as trees are harvested from the ground. Ladders are made of aluminum and are 3.7–4.0 m in length. Either four or six rows are harvested at a time, with an equal number of pickers distributed in each row as conditions warrant. Workers pick an entire tree and leapfrog one another down the rows. The foreman is responsible for moving the pickers between rows to maintain uniformity. Picking platforms have been tried in the past, but they are not an economically viable way of reducing reliance upon ladders due to their cost and the vast differences in trees and workers' efficiencies. When full, the bins are taken to a centralized area and unloaded from the bin-trailers or truck to await loading by forklift onto flatbed trailers for delivery to the packing facility. Full bins are typically covered with canvas to prevent heat damage, and loading areas are usually bordered by large shade trees that serve to help reduce fruit exposure to the sun. In instances where the orchard is close to the packing plant, the fruit can be conveyed there directly on the bin-trailers or truck. The fruit are hauled for short distances by trailers, but if the distance is longer than 10 km, the bins or totes are loaded on a truck for transportation to the packinghouses (Fig. 3.3).

Physical wounding of stone fruits can occur at any time from harvest until consumption, thus good worker supervision helps assure adequate

Fig. 3.3. Hauling peaches in totes. Photo courtesy of Dr. Carlos H. Crisosto.

protection against impact bruising and mechanical damage during picking, handling, and transport of fruit. Protection against roller bruising may require modifications of transport equipment, the time of day of harvest, and procedures. If severe injuries are encountered, consider using a top bin pad that maintains a slight tension across the top fruit. It is also helpful to grade farm roads to reduce roughness, avoid rough roads during transport, and establish strict speed limits for trucks operating between orchards and packinghouses. Stone fruits are transported from orchard to packinghouse and cooler as soon as possible after harvest. Fruit should be shaded during any delay between harvest and transport. Tractor drivers should be instructed to drive slowly and smoothly. Severe fruit damage can result from poor driving practices, especially on turns and starts. There are benefits to using "suspension-type" bin trailers instead of solid axle trailers. These trailers tend to ride more smoothly and induce less damage. Similar results can be obtained to a lesser degree by lowering tire air pressure. Unloading of trailers should also be performed as gently as possible. Care should be taken to educate workers as to the importance of this process. It is helpful if the unloading area is smooth and spacious to eliminate bumping and jarring. By choosing proper transportation routes and avoiding rough, bumpy roads fruit injury can be minimized. Position of fruit on the trailer is also important. Within-bin vibration levels are highest at the front of the trailer, intermediate in the rear, and lowest in the middle of the trailer (Crisosto and Costa, 2008).

Fig. 3.4. View of a general packing line showing peaches going through an automated sizing and grading operation. Photo courtesy of Dr. Carlos H. Crisosto.

Packinghouse Handling

Generally, peaches and nectarines are packed using automated packing lines (Fig. 3.4) and few are still hand packed. Most peaches and nectarines are partially hydrocooled or air cooled as soon as they arrive from the orchard.

Automatic-packing-line packed

At the packinghouse, some fruit are packed upon arrival from the orchard, others are partially cooled and packed the next day. In general, if fruit will not be packed within 2–3 days, they should be cooled close to 0°C to protect from deterioration. In all cases, it is highly recommended to cool peaches down to 0°C. At the packinghouse, the fruit are dumped (mostly using dry bin dumps) and cleaned using sanitation unit equipment where debris is removed and the fruit are sanitized. Peaches are normally washed using water containing chlorine and wet brushed to remove the trichomes, or fuzz, which are single cell extensions of epidermal cells (Fig. 3.5). Ideally, this area is located outside the packing area. After brushing-washing, fruit go through a short drying area in preparation for the waxing-fungicide application (Fig. 3.6) in the next protected area. Water-emulsified waxes are normally used, and approved fungicides may be incorporated into the wax (Fig. 3.7). Waxes are applied cold and heated drying is not necessary to spread and hold the applied fungicide as well as to add sheen and reduce moisture loss. Sorting or grading is done to

Fig. 3.5. Peach washing and brushing is the first step of the sanitation process. Photo courtesy of Dr. Carlos H. Crisosto.

Fig. 3.6. Peach drying after brushing and washing is the preparation for waxing and fungicide application. Photo courtesy of Dr. Carlos H. Crisosto.

Fig. 3.7. Peach waxing and fungicide application. Photo courtesy of Dr. Carlos H. Crisosto.

eliminate fruit with visual defects and sometimes to divert fruit of high surface color to a high-maturity pack (Fig. 3.8). Attention to detail is especially important for sorting line efficiency with peaches and nectarines, where a range of colors, sizes, and shapes of fruit can be encountered. Sizing segregates fruit by either weight or dimension and is carried out by operators or an electronic computer-controlled system (Fig. 3.9). Sorting and sizing equipment must be flexible to efficiently handle large volumes of small fruit or smaller volumes of larger fruit. Non-destructive sensors, such as near-infrared (NIR), are being used for quality segregation inline in a few packinghouses in the world. Most of the yellow-fleshed peaches and nectarines are packed in one-tray (flat) or two-tray boxes. In some cases, electronic weight sizers are used to automatically fill shipping containers (volume fill packed) with the fruit by weight. In other cases, mechanical place-packing units use hand-assisted fillers where the operator can control the belt speed to match the flow of fruit into plastic trays (Fig. 3.10). Most of the white-fleshed peaches and "tree-ripe" fruit are packed into one-tray boxes (flat) or clamshells (Fig. 3.11).

Ranch packed – high-maturity quality

A very limited volume of high-maturity stone fruits (soft) are "ranch packed" at the point of production. In these cases, the fruit are not washed, brushed, waxed, or treated with fungicide. Because they are handled less, a higher maturity index can be used, and growers can benefit from increased

Fig. 3.8. Initial grading and sorting of peaches. Photo courtesy of Dr. Carlos H. Crisosto.

Fig. 3.9. Automated sizing and sorting of peaches. Photo courtesy of Dr. Carlos H. Crisosto.

Fig. 3.10. Fruit packing into trays. Photo courtesy of Dr. Carlos H. Crisosto.

Fig. 3.11. Clamshell packing. Photo courtesy of Dr. Carlos H. Crisosto.

Fig. 3.12. Dumping buckets automatically into an automatic packing line for high-maturity quality peaches. Photo courtesy of Dr. Carlos H. Crisosto.

fruit size, red color, and high yield (Crisosto and Costa, 2008). These packers work directly from the buckets to select, grade, size, and pack the fruit into plastic trays. In Greece, most fruit are being harvested at the advanced maturity stage and directly hand packed at the orchard prior to being transferred to cold storage facilities.

In a high-maturity-quality ("tree-ripe") operation, which is becoming highly popular in the industry, high-maturity-quality fruit are picked, at an advanced maturity stage, into buckets or totes that are carried by trailer to the automatic packing line area (Fig. 3.12). Then, fruit are gently dumped into smooth packing lines for washing, brushing, waxing, sorting, and packaging. A large volume of the Californian industry is using high-quality fruit that can be produced by managing the orchard factors (fruit thinning, girdling, fertilization, irrigation) properly and picking firm fruit (Crisosto and Costa, 2008). In this case, preconditioning at the shipping is essential to assure high flavor quality for consumers. If a preconditioning protocol is not applied, which is ideal to protect consumers, ripening should be carried out at the retail stores prior to sales.

Chilling Injury (CI)

The major physiological cause of deterioration in peach and nectarine is the incidence of physiological disorders due to extended low temperature, evident as CI (Lurie and Crisosto, 2005; Martínez-García *et al.*, 2012;

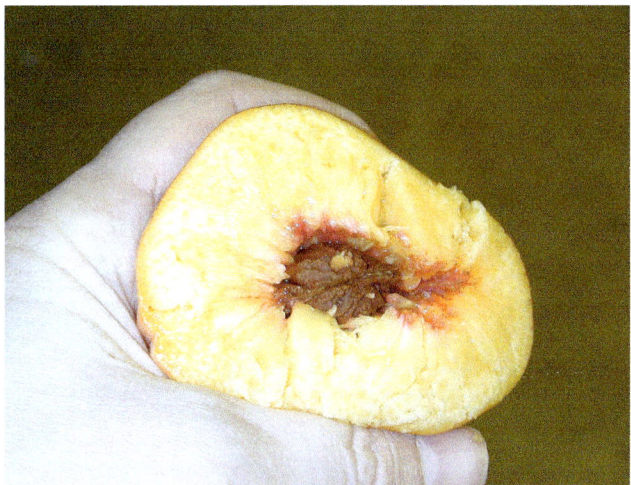

Fig. 3.13. Mealy (M) peach, a chilling injury (CI) symptom. Photo courtesy of Dr. Carlos H. Crisosto.

Manganaris *et al.*, 2019). The disorder can manifest itself as dry, mealy (M), woolly (Fig. 3.13), or hard-textured fruit with lack of juice (Crisosto and Labavitch, 2002), or with flesh browning (FB) (Manganaris *et al.*, 2007) or pit cavity browning that usually radiates through the flesh from the pit (Crisosto *et al.*, 1999; Brummell *et al.*, 2004b). An intense red bleeding color in the flesh (Fbl, Fig. 3.14), usually radiating from the pit, may be associated with this CI problem in some cultivars (Lurie and Crisosto, 2005;

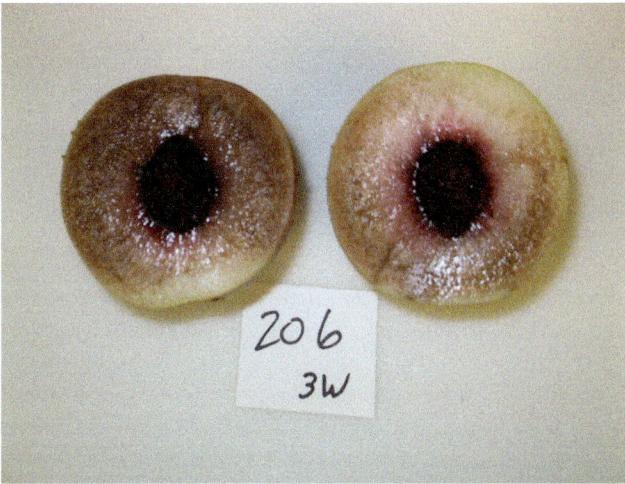

Fig. 3.14. Flesh bleeding (Fbl) and flesh browning (FB), as CI symptoms. Photo courtesy of Dr. Carlos H. Crisosto.

Fig. 3.15. Juicy peach showing no incidence of CI. Photo courtesy of Dr. Carlos H. Crisosto.

Manganaris *et al.*, 2008). In most cases, development of M texture, changes from juicy (Fig. 3.15) to dry, occurs prior to FB. However, in all freestone peaches, flavor is lost before visual symptoms are evident (Crisosto, 2002). The evolution of such symptoms makes it very difficult to deal with the problems during postharvest handling as fruit can be sensory damaged before visual symptoms develop and are detected by an inspector or consumers (Crisosto *et al.*, 1999; Crisosto and Tonutti, 2015). There is large variability in CI susceptibility among peach and nectarine cultivars that indicated that the problem has genetic origin (Martínez-García *et al.*, 2012, reviewed in Manganaris *et al.*, 2019). In general, peach cultivars are more susceptible than nectarine (Crisosto *et al.*, 1999, 2009; Crisosto and Valero, 2008). In susceptible cultivars, CI symptoms develop faster and more intensely when fruit are stored at temperatures between 2°C and 7°C than when fruit are stored at 0°C or below, but above the tissue's freezing point. Thus, the temperatures between 2°C and 7°C are called the "killing temperature zone" (Crisosto *et al.*, 1999; Lurie and Crisosto, 2005; Manganaris *et al.*, 2019). At the shipping point, fruit should be cooled and held near or below 0°C. During transportation, holding CI-susceptible cultivars at 5°C will significantly reduce their postharvest life. For example, the market life of popular freestone, melting yellow-fleshed cultivars susceptible to CI, such as 'Elegant Lady' and 'O'Henry,' is reduced from 5 weeks at 0°C down to 1–2 weeks when fruit is handled at 5°C. Current commercial cultivars growing in Europe exhibited long postharvest life when stored at 5°C and 0°C.

Transportation

CI-susceptible cultivars that are exposed to approximately 5°C can have a significantly reduced postharvest life. Many treatments have been tested to delay and limit development of CI. The success of controlled atmosphere (CA) treatment (6 kPa O_2 with 17 kPa CO_2) in ameliorating CI, especially FB, depends on cultivar market life, fruit maturity, shipping time, and fruit size (Lurie and Crisosto, 2005; Crisosto *et al.*, 2009; Manganaris and Crisosto, 2020). A controlled delayed cooling or preconditioning treatment can extend the market life and is being successfully and commercially used in the USA and other countries. A 48 h cooling delay at 20°C (preconditioning treatment) and high relative humidity can extend the market life of CI-susceptible peaches by up to 2 weeks without causing fruit deterioration (Crisosto *et al.*, 2004b). Breeding programs are starting to use marker assisted selection technology to identify genes related to CI in order to avoid using these in their crosses (Peace *et al.*, 2005; Pons *et al.*, 2014, 2015; Fresnedo-Ramírez *et al.*, 2015).

Physical Damage

Stone fruits are susceptible to mechanical injuries including impact, compression, and abrasion (or vibration) bruising. Careful handling during harvest, handling, and packing operations is important because such injuries result in reduced appearance, accelerated physiological activity, potentially more inoculation by fruit decay organisms, and greater water loss (Crisosto and Costa, 2008). Incidence of impact (Fig. 3.16) and compression bruising has become a greater concern as a large part of our industry is harvesting fruit at high maturity, and therefore when its flesh is softer, to maximize fruit orchard quality. Several packing line surveys that have been carried out indicated that most impact bruising damage occurs during the packinghouse operation and long transportation from the orchard to the packinghouse. Critical impact bruising thresholds, the minimum fruit firmness measured at the weakest point to tolerate impact abuse, have been developed for three potential bruising packinghouse conditions for many peach, nectarine, and plum cultivars (Crisosto *et al.*, 2001b, 2004a). Practical recommendations to reduce physical abuse in the packinghouse have been published (Crisosto *et al.*, 2001b; Crisosto and Costa, 2008).

Abrasion damage

This occurs at any time during postharvest handling (Crisosto and Costa, 2008). Protection against abrasion damage involves procedures to reduce vibrations during transport and handling by immobilizing the fruit. Steps can include: (i) installing air suspension systems on axles of field and highway trucks; (ii) using plastic sideliners inside field bins; (iii)

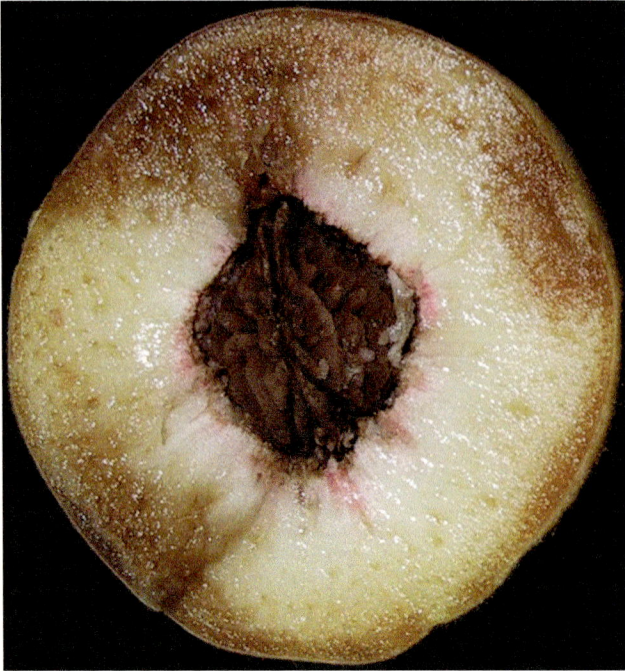

Fig. 3.16. Bruises caused by impact damage during harvesting-packing operations on a white-fleshed peach. Photo courtesy of Dr. Carlos H. Crisosto.

installing special bin-top pads before transport; (iv) avoiding abrasion on the packing line; and (v) using packing procedures that immobilize the fruit within the shipping container before they are transported to market. Using medium-sized plastic bins has also reduced abrasion and impact damage.

Skin inking or field skin discoloration

In situations when abrasion damage occurred during harvesting of fruit that have heavy metal contaminants such as iron, copper and/or aluminum on their skin, a dark discoloration (Fig. 3.17) is formed on the surface of peaches and nectarines (Cheng and Crisosto, 1997; Serrano *et al.*, 2004; Crisosto and Costa, 2008). These dark-colored or brown spots or stripes on the fruit, inking or peach skin discoloration, become a cosmetic problem and a reason for discard. Heavy metal contaminants on the surface of the fruit may occur due to the spraying of foliar nutrients less than 15 days before harvest or the spraying of fungicides less than 7 days before harvest. Preharvest intervals have been developed for several approved fungicides

Fig. 3.17. Peach and nectarine inking or skin discoloration. Photo courtesy of Dr. Carlos H. Crisosto.

in California. Light-brown spots or stripes are also produced on the surface of white-fleshed peaches and nectarines because of abrasion occurring mainly during harvest and hauling operations.

Skin browning or packinghouse skin discoloration-burning

A skin discoloration damage symptom, evident as burning, has been reported in white-fleshed cultivars (Cantín *et al.*, 2011). These brown spots or stripes only developed during packinghouse operations and were related to high pH exposures during the sanitation operation and were restricted to a specific group of cultivars. Controlling the pH during chlorine application that also improves sanitation eliminated the problem.

Fruit shriveling

Loss of approximately 5–8% of water, based on weight at harvest, may cause visual shrivel in peaches and nectarines. While there is a large variability in susceptibility to water loss among cultivars, all stone fruits must be protected to assure their postharvest performance. Mineral oil waxes can potentially control water losses better than vegetable oil and edible coatings. Because fruit shrivel results from cumulative water loss throughout handling, it is important to maintain low temperature and high relative humidity throughout harvesting, packing, storage, transport, and distribution.

Short cooling delays, efficient waxing and brushing, fast cooling followed by storage under constant low temperature and high relative humidity are the main ways of limiting water loss.

Temperature Management

The ideal peach storage temperature is $-1°C$ to $0°C$. The flesh freezing point varies depending on the total soluble solids (TSS). Storage-room relative humidity should be maintained at 90–95% and an airflow of approximately 0.0236 m^3 s^{-1} t^{-1} is suggested during storage. Cooling requirements depend in part upon the scheduling of the packing operation. At arrival in the packinghouse, fruit can be cooled in field bins using forced-air cooling or hydrocooling.

Some peaches and nectarines are hydrocooled using a conveyor-type hydrocooler while others are hydrocooled *in situ*. However, cooling of packed fruit is carried out by forced-air systems. On arrival at the packinghouse, fruit in field bins can be cooled to intermediate temperatures (between 5°C and 10°C) provided packing will occur the next day. If packing is to be delayed beyond the next day, then fruit should be thoroughly cooled in the bins to near 0°C. In CI-susceptible and fast-softening cultivars, short cooling delays (within 8 h of harvest), fast cooling and maintaining fruit temperature near 0°C is recommended. Fruit in packed containers should be cooled to near 0°C but maintaining such low temperatures requires knowledge of the freezing point of the fruit and of the temperature fluctuations in the storage system (Fig. 3.18). Even peaches that were thoroughly cooled in the bins will warm substantially during packing and should be thoroughly re-cooled after packing.

Shipping and transportation

Stone fruit storage and overseas shipments should be at or below 0°C. The temperature during truck transportation within the USA, Canada, and Mexico should be below 2.2°C. Holding stone fruits at these low temperatures minimizes the losses associated with rotting organisms, excessive softening, water losses, and the deterioration resulting from CI in susceptible cultivars (Crisosto *et al.*, 1999). At the shipping point, fruit should be cooled and held near or below 0°C, in accordance with their freezing point. Peach storage and overseas shipments should be at or below 0°C. Maintaining these low-pulp temperatures requires knowledge of the freezing point of the fruit, the temperature fluctuations in the storage system, loading techniques, and equipment performance. Holding peaches at these low temperatures minimizes the losses associated with rotting organisms, excessive softening and water losses, and the deterioration resulting

Fig. 3.18. Typical Californian peach cold storage room. Photo courtesy of Dr. Carlos H. Crisosto.

from CI in susceptible cultivars, therefore optimizing their postharvest life (Crisosto and Valero, 2008).

Special Storage Treatments

Controlled atmosphere (CA)

The major benefits of CA for yellow-fleshed cultivars and white-fleshed cultivars during storage and shipment are retention of fruit firmness and ground color. Hence a CA of $6\,kPa$ O_2 and $17\,kPa$ CO_2 at $0°C$ has been recommended for retention of fruit firmness, delaying ground color changes, and reduction of FB during shipping. However, mealiness and "off-flavor" were not affected. In a detailed study testing four CA conditions (air, $2\,kPa$ O_2 and $5\,kPa$ CO_2, $10\,kPa$ CO_2 and $10\,kPa$ O_2, and $17\,kPa$ CO_2 and $6\,kPa$ O_2) for 5 and 6 weeks (simulated shipment) on 11 important nectarine cultivars grown in California, it was concluded that in nine out of the 11 CA was not necessary after 5 weeks storage at $0°C$ as CI symptoms were not expressed and fruits remained firm. In detailed CA evaluation studies, in which fruits were held in cold storage for 1 week before CA treatments were established, simulating standard shipping practices, 'August Pearl' had high FB incidence that was reduced by any of the three CA treatments. By 5 weeks of simulated shipment, 'August Fire' had high levels of mealiness in all the treatments. On the other hand, by 6 weeks of simulated shipment at

0°C, only six cultivars out of the 11 did not show commercial CI symptoms in any of the storage conditions. 'Ruby Diamond' had near 80% FB and absence of flesh mealiness in fruit under cold storage and FB was reduced in all the CA treatments. 'Honey Royale', 'Summer Bright', 'August Pearl', and 'August Fire' mainly had a high incidence of flesh mealiness that was not reduced by using any of the CA treatments. Despite the lack of good control of flesh mealiness, the CA treatments did not damage fruit except for the 17 kPa CO_2 and 6 kPa O_2 treatment. A dull skin browning color and pitting was observed on 'August Pearl' and 'Honey Royale' fruit after 6 weeks cold storage under the 6 kPa O_2 and 17 kPa CO_2 treatment (Crisosto et al., 2009; Manganaris and Crisosto, 2020). The efficacy of CA treatment is related to cultivar, preharvest factors, fruit size, temperature, market life, and shipping time (Crisosto et al., 2009; Manganaris and Crisosto, 2020). For example, small size peaches stored in air at 0°C have a longer market life than large fruit. At 0°C and 5°C, large-sized 'Elegant Lady' and 'O'Henry' fruit have a longer market life under CA than under air storage. In California clingstone canning peaches, a large difference in storage potential and canned quality following storage was reported among five clingstone peach cultivars tested (Crisosto et al., 2009). Based on this study, it is recommended that 'Loadel' and 'Carolyn' canning peaches should be stored for up to 4 weeks under 2 KPa O_2 and 5 KPa CO_2 at 1.2°C while 'Andross' and 'Halford' peaches should be stored for a shorter storage period. In most cultivars, visual development of FB occurred later than loss of flavor, "off-flavor" and mealiness development that are the limitation to peach and nectarine consumption. Therefore, the use of CA technology during storage and shipment has been limited (Crisosto et al., 2009; Crisosto and Tonutti, 2015).

Modified atmosphere packaging (MAP)

This technique has been tested on several peach and nectarine cultivars without success under laboratory and Californian handling-marketing conditions (Manganaris and Crisosto, 2020). Despite high CO_2 levels that were reached inside the liner during cold storage, flesh mealiness and FB development limited the potential benefits of this technology. In some commercial cases when box liners (MAP) were used the incidence of decay and "off-flavor" increased due to lack of proper cooling and condensation during transportation (Crisosto et al., 2009). The use of current MAP technologies has very limited benefits on maintaining peach or nectarine postharvest life.

Preconditioning or controlled delayed cooling

A successful controlled delayed cooling protocol for commercial use was developed to reduce CI, extend the postharvest life of peaches and nectarines and to deliver fruit to consumers close to the "ready-to-eat" stage. During this controlled partial ripening, fruit CI sensitivity decreases, and fruit reach the "ready-to-buy" or "ready-to-transfer" stage. The application of this treatment at the shipping point is critical as CI damage is triggered during shipment and handling at the retail store. Thus, when this preconditioning treatment is properly applied, peaches and nectarines are protected against CI injury and are at the "ready-to-buy" and close to the "ready-to-eat" stage that increase consumer acceptance. This treatment has become very popular in the last few years in California and other countries due to high consumer satisfaction. Currently, several stone fruit companies employ this process using large ripening rooms. A 48 h cooling delay at 20°C applied before precooling is the most effective treatment for extending market life of CI-susceptible peaches without causing fruit deterioration. This treatment increased the minimum market life by up to 2 weeks in the cultivars tested when exposed to the "killing temperature zone". Minimal weight loss and softening occurred during the controlled delayed cooling treatments without reducing fruit quality. However, fruit must be cooled down and fruit temperature should be ideally maintained near 0°C during their postharvest handling.

Pathological Problems

Postharvest loss of stone fruits due to decay-causing fungi is considered the greatest deterioration problem. Worldwide, the most common pathogen of fresh stone fruits is gray mold, caused by the fungus *Botrytis cinerea* (Crisosto and Valero, 2008). In California, an even greater cause of loss is brown rot (Fig. 3.19), caused by the fungus *Monilinia fructicola*, *Rhizopus* rot (caused by *Rhizopus stolonifer*) or sour rot (Fig. 3.20) (caused by *Geotrichum candidum* complex) which can occur in ripe or near-ripe peaches kept at 20–25°C (Adaskaveg *et al.*, 2008). Cooling and keeping fruit below 5°C are part of an effective control. Good orchard sanitation practices and fungicide applications are essential to reduce these problems (Adaskaveg *et al.*, 2008). It is common to use a postharvest fungicidal treatment against these diseases. A US Environmental Protection Agency (EPA)-approved fungicide is often incorporated into a fruit wax for uniformity of application. Careful handling to minimize fruit injury, sanitation of packinghouse equipment, and rapid, thorough cooling to 0°C as soon after harvest as possible are also important for effective disease suppression.

Suitability as a Fresh-cut Product

Very limited peach fresh-cut marketing has been developed because of the short market life of this produce reported in early studies (Gorny *et al.*, 1999; Crisosto and Tonutti, 2015). The optimal ripeness for preparing

Fig. 3.19. Peach brown rot (caused by *Monilinia fructicola*). Photo courtesy of Dr. Carlos H. Crisosto.

Fig. 3.20. Peach sour rot (caused by *Geotrichum candidum* complex). Photo courtesy of Dr. Carlos H. Crisosto.

fresh-cut peach slices is when the flesh firmness reaches 1.4–2.7 kgf, and these slices can retain good eating quality for 2–8 days (depending on cultivar) while kept at 5°C and 90–95% relative humidity. Post-slice dips in ascorbic acid and calcium lactate or use of MAP may slightly prolong the shelf life of peach slices. Recently, mild heat pretreatments (40°C for 70 min) before minimal processing and packing under passive MAP conditions were effective in inducing firmness, while preserving nutritional quality (Steiner *et al.*, 2006).

Peach Handling at Retail Stores

Peaches should be transported at 0–1.7°C from the distribution center and kept at this temperature range prior to transfer to a dry/warm table for display. In situations where fruit temperature cannot be maintained out of the "killing temperature zone", it would be preferable to move fruit fast. Firmness measurements need to be considered in the decision-making process. Peaches should ideally be arriving from the distribution center to the retail stores with firmness in the range of 1.8–2.7 kgf (weakest position) or 2.7–3.6 kgf (cheeks). This fruit is at the "ready-to-buy" or "transfer-point" stage of ripening and within ~48–72 h at 20°C should be "ready-to-eat" in the 0.9–1.8 kgf firmness range. This is the firmness range at which most consumers claim the highest satisfaction when eating peaches. Produce managers need to be educated about this new "ready-to-buy" type of fruit (preconditioned) to minimize mechanical damage and expedite an effective rotation (first in, first out). Peaches should be displayed on dry tables and labelled well as "ready-to-buy/eat", and consumers should understand that this fruit is riper than conventionally packed tree fruit. In order to protect these peaches, the display should be no more than two layers deep and in-box display should be attempted. As tree fruit will continue to ripen on the display warm/dry table, they should be checked often, and the softest fruit be placed at the front of the display. Fruit that reach the "ready-to-eat" ripeness (firmness: 0.9–1.4 kgf) need to be sold quickly or refrigerated to extend their shelf life. It is essential that consumers be instructed that this type of fruit should be refrigerated if it is not going to be consumed within 3 days of purchase.

Cull Utilization

The main use of peach culls is for cattle feed because culled peach is palatable and a good source of energy (Crisosto and Valero, 2008), but it is low in protein and has other characteristics that make it different from other feed sources. For example, peaches contain ~85% water, 9% digestible dry matter, 5% pits and 2% indigestible dry matter. The high-water content diminishes the real value as feed because it makes culls expensive to transport, requires large trough volumes, and allows the feed to spoil quickly. If fed in large proportions, culled fruit causes almost continuous urination and consequently the animals require a high amount of salt. The only potential advantage to the high-water content is that animals in a remote, dry location will not need extra water hauled to them. Low protein levels in culled fruit limit the quantity that can be fed where rapid weight gain is important, such as in feedlots. The use of culls for fuel alcohol production is limited mainly by the low sugar content; thus, peach is not included in this group. The 8–12% sugar content of most culled peaches results in an alcohol yield of about 421 t^{-1} of fruit, which is too low compared with potatoes (83–1041 t^{-1}) or maize (3751 t^{-1}). This low yield makes it uneconomical in addition to the waste management problem (Thompson, 2002). Unfortunately, the limits to the use of culls often result in large portions of them being discarded. In general, peach culls are going for frozen or canned peaches or juice, dried for charity donation, or used for livestock feed. When fruit have worms and decay, they are utilized as green waste for compost. The amount of culls varies according to season, cultivar and other conditions from ~10% to 30% of total production. The decision on the cull utilization is made based on returns. In general, when the reason for disposal has been small sizes and mainly cosmetic blemishes, fruit still have value for human consumption and can be frozen, canned or used for making juice or other value-added products. Improper disposal can cause sanitary and pollution problems. Flies and odor problems can be prevented by ensuring rapid drying. Fly maggots hatch into adults within 7–10 days, and odor problems can develop before flies appear. The culls should be crushed and spread no more than one or two layers deep; sometimes this is done on orchard roads or fallow fields. Culls can be disked into the soil, although this tends to cover the fruit with soil and slows drying; also, insects or diseases that may have caused the fruit to be culled in the first place may infect a future crop. Disposal sites should be as far away from neighbors as possible. Flies can travel up to 8 km from the place where they hatch. Culls should not be dumped near streambeds. Fruit cull piles can attract the dumping of many other kinds of refuse. If culls are deposited away from the point of production, use municipal solid waste disposal sites if available. Some culls can be turned into dried fruit for human consumption. However, good-quality dried fruit is made only from good-quality fresh fruit. Only undersized or slightly overripe fruit should be considered for drying.

Special Research Needs

Research needs include:

- removing CI susceptibility by using a classic breeding program (marker assisted selection) and/or gene manipulation;
- selecting canopy architecture for adaptation to mechanical harvesting;

- improving flavor diversity in the available peach and nectarine cultivars; and
- developing low chilling, flavorful cultivars with a postharvest life to keep up with global warming and new attractive production areas to extend year-around availability.

Acknowledgments

University of California Agriculture and Natural Resources (UC ANR) retains the copyright of the short version of this chapter on peach and nectarine that will be published in *Postharvest Technology of Horticulture Crops*, 4th edn and CABI will hold the copyright for this long version.

References

Adaskaveg, J.E., Schnabel, G. and Foster, H. (2008) Diseases of peach caused by fungi and fungal-like organisms: biology, epidemiology, and management. In: Layne, D.R. and Bassi, D. (eds) *The Peach: Botany, Production and Uses*. CAB International, Wallingford, UK, pp. 352–407.

Bonany, J., Carbó, J., Echeverria, G., Hilaire, C., Cottet, V. *et al.* (2014) Eating quality and European consumer acceptance of different peach (*Prunus persica* (L.) Batsch) varieties. *Journal of Food, Agriculture & Environment* 12, 67–72.

Brummell, D.A., Cin, V.D., Crisosto, C.H. and Labavitch, J.M. (2004a) Cell wall metabolism during maturation, ripening and senescence of peach fruit. *Journal of Experimental Botany* 55(405), 2029–2039. DOI: 10.1093/jxb/erh227.

Brummell, D.A., Cin, V.D., Lurie, S., Crisosto, C.H. and Labavitch, J.M. (2004b) Cell wall metabolism during the development of chilling injury in cold-stored peach fruit: association of mealiness with arrested disassembly of cell wall pectins. *Journal of Experimental Botany* 55(405), 2041–2052. DOI: 10.1093/jxb/erh228.

Byrne, D.H., Noratto, G., Cisneros-Zevallos, L., Porter, W. and Vizzotto, M. (2009) Health benefits of peach, nectarine and plums. *Acta Horticulturae* 841(841), 267–274. DOI: 10.17660/ActaHortic.2009.841.32.

Cantín, C.M., Tian, L., Qin, X. Crisosto, C.H. and Xiaoqiong, Q. (2011) Copigmentation triggers the development of skin burning disorder on peach and nectarine fruit [*Prunus persica* (L.) Batsch]. *Journal of Agricultural and Food Chemistry* 59(6), 2393–2402. DOI: 10.1021/jf104497s.

Cheng, G.W. and Crisosto, C.H. (1997) Iron–polyphenol complex formation and skin discoloration in peaches and nectarines. *Journal of the American Society for Horticultural Science* 122(1), 95–99. DOI: 10.21273/JASHS.122.1.95.

Crisosto, C.H. (2002) How do we increase peach consumption? *Acta Horticulturae* 592(592), 601–605. DOI: 10.17660/ActaHortic.2002.592.82.

Crisosto, C.H. and Costa, G. (2008) Preharvest factors affecting peach quality. In: Layne, D.R. and Bassi, D. (eds) *The Peach: Botany, Production and Uses*. CAB International, Wallingford, UK, pp. 536–549.

Crisosto, C.H. and Crisosto, G.M. (2005) Relationship between ripe soluble solids concentration (RSSC) and consumer acceptance of high and low acid melting flesh peach and nectarine (*Prunus persica* (L.) Batsch) cultivars. *Postharvest Biology and Technology* 38(3), 239–246. DOI: 10.1016/j.postharvbio.2005.07.007.

Crisosto, C.H. and Labavitch, J.M. (2002) Developing a quantitative method to evaluate peach (*Prunus persica*) flesh mealiness. *Postharvest Biology and Technology* 25(2), 151–158. DOI: 10.1016/S0925-5214(01)00183-1.

Crisosto, C.H. and Tonutti, P. (2015) Innovations in peach postharvest research and storage technology. *Acta Horticulturae* 1084(1084), 821–828. DOI: 10.17660/ActaHortic.2015.1084.111.

Crisosto, C.H. and Valero, D. (2008) Harvesting and postharvest handling of peaches for the fresh market. In: Layne. and Bassi. (eds) *The Peach: Botany, Production and Uses.* CAB International, pp. 575–596.

Crisosto, C.H., Mitchell, F.G. and Ju, Z. (1999) Susceptibility to chilling injury of peach, nectarine, and plum cultivars grown in California. *HortScience* 34(6), 1116–1118. DOI: 10.21273/HORTSCI.34.6.1116.

Crisosto, C.H., Day, K.R., Crisosto, G.M. and Garner, D. (2001a) Quality attributes of white flesh peaches and nectarines grown under California conditions. *Journal of the American Pomological Society* 55, 45–51.

Crisosto, C.H., Slaughter, D., Garner, D. and Boyd, J. (2001b) Stone fruit critical bruising thresholds. *Journal of the American Pomological Society* 55, 76–81.

Crisosto, C.H., Slaughter, D. and Garner, D. (2004a) Developing maximum maturity indices for 'Tree ripe' fruit. *Advances in Horticultural Science* 18, 29–32.

Crisosto, C.H., Garner, D., Andris, H.L. and Day, K.R. (2004b) Controlled delayed cooling extends peach market life. *HortTechnology* 14(1), 99–104. DOI: 10.21273/HORTTECH.14.1.0099.

Crisosto, C.H., Crisosto, G.M., Echeverria, G. and Puy, J. (2006) Segregation of peach and nectarine (*Prunus persica* (L.) Batsch) cultivars according to their organoleptic characteristics. *Postharvest Biology and Technology* 39(1), 10–18. DOI: 10.1016/j.postharvbio.2005.09.007.

Crisosto, C.H., Lurie, S. and Retamales, J. (2009) Stone fruit. In: Yahia, E.M. (ed.) *Modified and Controlled Atmospheres for the Storage, Transportation, and Packaging of Horticultural Commodities.* CRC Press, Boca Raton, Florida, pp. 287–315.

Delgado, C., Crisosto, G.M., Heymann, H. and Crisosto, C.H. (2013) Determining the primary drivers of liking to predict consumers' acceptance of fresh nectarines and peaches. *Journal of Food Science* 78(4), S605–S614. DOI: 10.1111/1750-3841.12063.

Drogoudi, P., Pantelidis, G.E., Goulas, V., Manganaris, G.A., Ziogas, V. *et al.* (2016) The appraisal of qualitative parameters and antioxidant contents during postharvest peach fruit ripening underlines the genotype significance. *Postharvest Biology and Technology* 115, 142–150. DOI: 10.1016/j.postharvbio.2015.12.002.

FAO (2017) FAOSTAT Database. Food and Agriculture Organization of the United Nations, Rome. Available at: http://www.fao.org/statistics/en/ (accessed 23 January 2020).

Ferrer, A., Remón, S., Negueruela, A.I. and Oria, R. (2005) Changes during the ripening of the very late season Spanish peach cultivar Calanda. Feasibility of using CIELAB coordinates as maturity indices. *Scientia Horticulturae* 105, 435–446.

Fresnedo-Ramírez, J., Bink, M.C.A.M., van de Weg, E., Famula, T.R., Crisosto, C.H. *et al.* (2015) QTL mapping of pomological traits in peach and related species breeding germplasm. *Molecular Breeding* 35(8). DOI: 10.1007/s11032-015-0357-7.

Gil, M.I., Tomás-Barberán, F.A., Hess-Pierce, B. and Kader, A.A. (2002) Antioxidant capacities, phenolic compounds, carotenoids, and vitamin C contents of nectarine, peach, and plum cultivars from California. *Journal of Agricultural and Food Chemistry* 50(17), 4976–4982. DOI: 10.1021/jf020136b.

Goristein, S., Martín-Belooso, O., Lojek, A., Ciz, M., Soliva-Fortuny, R. *et al.* (2002) Comparative content of some phytochemicals in Spanish apples, peaches and pears. *Journal of the Science of Food and Agriculture* 82(10), 1166–1170. DOI: 10.1002/jsfa.1178.

Gorny, J.R., Hess-Pierce, B. and Kader, A.A. (1999) Quality changes in fresh-cut peach and nectarine slices as affected by cultivar, storage atmosphere and chemical treatments. *Journal of Food Science* 64(3), 429–432. DOI: 10.1111/j.1365-2621.1999.tb15057.x.

Iglesias, I. (2019) Peach production in Spain: current situation and trends, from production to consumption. In: Milatovic, D. (ed.) *Proceedings of the 4th Conference, Innovations in Fruit Growing*. Faculty of Agriculture, Belgrade (Serbia), pp. 75–94.

Iglesias, I. and Echeverría, G. (2009) Differential effect of cultivar and harvest date on nectarine colour, quality and consumer acceptance. *Scientia Horticulturae* 120(1), 41–50. DOI: 10.1016/j.scienta.2008.09.011.

Kahlon, T.S. and Smith, G.E. (2007) *In vitro* binding of bile acids by bananas, peaches, pineapple, grapes, pears, apricots and nectarines. *Food Chemistry* 101(3), 1046–1051. DOI: 10.1016/j.foodchem.2006.02.059.

Lavilla, T., Recasens, I., López, M.L. and Puy, J. (2002) Multivariate analysis of maturity stages, including quality and aroma, in 'Royal Glory' peaches and 'Big Top' nectarines. *Journal of the Science of Food and Agriculture* 82(15), 1842–1849. DOI: 10.1002/jsfa.1268.

Lee, C.K., Park, K.-K., Hwang, J.-K., Lee, S.K. and Chung, W.-Y. (2008) The pericarp extract of *Prunus persica* attenuates chemotherapy-induced acute nephrotoxicity and hepatotoxicity in mice. *Journal of Medicinal Food* 11(2), 302–306. DOI: 10.1089/jmf.2007.545.

Lurie, S. and Crisosto, C.H. (2005) Chilling injury in peach and nectarine. *Postharvest Biology and Technology* 37(3), 195–208. DOI: 10.1016/j.postharvbio.2005.04.012.

Manganaris, G.A., Vasilakakis, M., Diamantidis, G. and Mignani, I. (2007) The effect of postharvest calcium application on tissue calcium concentration, quality attributes, incidence of flesh browning and cell wall physicochemical aspects of peach fruits. *Food Chemistry* 100(4), 1385–1392. DOI: 10.1016/j.foodchem.2005.11.036.

Manganaris, G.A., Vasilakakis, M., Mignani, I. and Manganaris, A. (2008) Cell wall physicochemical properties as indicators of peach quality during fruit ripening after cold storage. *Food Science and Technology International* 14(4), 385–391. DOI: 10.1177/1082013208097251.

Manganaris, G.A., Drogoudi, P., Goulas, V., Tanou, G., Georgiadou, E.C. *et al.* (2017) Deciphering the interplay among genotype, maturity stage and low-temperature storage on phytochemical composition and transcript levels of enzymatic antioxidants in *Prunus persica* fruit. *Plant Physiology and Biochemistry* 119, 189–199. DOI: 10.1016/j.plaphy.2017.08.022.

Manganaris, G.A., Vincente, A.R., Martinez, P. and Crisosto, C.H. (2019) Postharvest physiological disorders in peach and nectarine. In: Tonetto de Freitas, S. and Pareek, S. (eds) *Physiological Disorders in Fruits and Vegetables.* 9781138035508. CRC press, pp. 253–264.

Manganaris, G.A. and Crisosto, C.H. (2020) Stone fruits. In: Gil, M. and Beaudry, R. (eds) *Controlled and Modified Atmosphere for Fresh and Fresh-Cut Produce.* Elsevier, Amsterdam.

Martínez-García, P.J., Peace, C.P., Parfitt, D.E., Ogundiwin, E.A., Fresnedo-Ramírez, J. *et al.* (2012) Influence of year and genetic factors on chilling injury susceptibility in peach (*Prunus persica* (L.) Batsch). *Euphytica* 185(2), 267–280. DOI: 10.1007/s10681-011-0572-1.

Moriguchi, T., Ishizawa, Y. and Sanada, T. (1990) Differences in sugar composition in *Prunus persica* fruit and the classification by the principal component analysis. *Journal of the Japanese Society for Horticultural Science* 59(2), 307–312. DOI: 10.2503/jjshs.59.307.

Noratto, G., Porter, W., Byrne, D. and Cisneros-Zevallos, L. (2009) Identifying peach and plum polyphenols with chemopreventive potential against estrogen-independent breast cancer cells. *Journal of Agricultural and Food Chemistry* 57(12), 5219–5226. DOI: 10.1021/jf900259m.

Noratto, G., Porter, W., Byrne, D. and Cisneros-Zevallos, L. (2014) Polyphenolics from peach (*Prunus persica* var. Rich Lady) inhibit tumor growth and metastasis of MDA-MB-435 breast cancer cells *in vivo. The Journal of Nutritional Biochemistry* 25(7), 796–800. DOI: 10.1016/j.jnutbio.2014.03.001.

Peace, C.P., Crisosto, C.H. and Gradziel, T.M. (2005) Endopolygalacturonase: a candidate gene for Freestone and melting Fleshin peach. *Molecular Breeding* 16(1), 21–31. DOI: 10.1007/s11032-005-0828-3.

Pons, C., Martí, C., Forment, J., Crisosto, C.H., Dandekar, A.M. *et al.* (2014) A bulk segregant gene expression analysis of a peach population reveals components of the underlying mechanism of the fruit cold response. *PLoS ONE* 9(3), e90706. DOI: 10.1371/journal.pone.0090706.

Pons, C., Dagar, A., Marti Ibanez, C., Singh, V., Crisosto, C.H. *et al.* (2015) Pre-Symptomatic transcriptome changes during cold storage of chilling sensitive and resistant peach cultivars to elucidate chilling injury mechanisms. *BMC Genomics* 16, 245.

Proteggente, A.R., Pannala, A.S., Paganga, G., Van Buren, L., Wagner, E. *et al.* (2002) The antioxidant activity of regularly consumed fruit and vegetables reflects their phenolic and vitamin C composition. *Free Radical Research* 36(2), 217–233. DOI: 10.1080/10715760290006484.

Serrano, M., Martínez-Romero, D., Castillo, S., Guillén, F. and Valero, D. (2004) Effect of preharvest sprays containing calcium, magnesium and titanium on the quality of peaches and nectarines at harvest and during postharvest storage. *Journal of the Science of Food and Agriculture* 84(11), 1270–1276. DOI: 10.1002/jsfa.1753.

Steiner, A., Abreu, M., Correia, L., Beirão-da-Costa, S., Leitão, E. *et al.* (2006) Metabolic response to combined mild heat pre-treatments and modified atmosphere packaging on fresh-cut peach. *European Food Research and Technology* 222(3-4), 217–222. DOI: 10.1007/s00217-005-0025-y.

Thompson, J.F. (2002) Cull utilization. In: Kader, A.A. (ed.) *Postharvest Technology of Horticultural Crops. Special Publication No. 3311.* University of California, Division of Agriculture and Natural Resources, Oakland, California, pp. 41–43.

Tomás-Barberán, F.A., Gil, M.I., Cremin, P., Waterhouse, A.L., Hess-Pierce, B. *et al.* (2001) HPLC-DAD-ESIMS analysis of phenolic compounds in nectarines, peaches, and plums. *Journal of Agricultural and Food Chemistry* 49(10), 4748–4760. DOI: 10.1021/jf0104681.

USDA (2018) USDA Agricultural Research Service (ARS) Food Data Central Peach Nutrients Search Results. US Department of Agriculture, Washington, DC. Available at: https://fdc.nal.usda.gov/fdc-app.html#/food-details/438781/nutrients (accessed 5 December 2019).

Vicente, A.R., Manganaris, G.A., Sozzi, G.O. and Crisosto, C.H. (2009) Nutritional quality of fruits and vegetables. In: Florkowski, W., Shewfelt, R., Brueckner, B. and Prussia, S. (eds) *Postharvest Handling: a Systems Approach.* Academic Press, Elsevier, Oxford, pp. 57–106.

Vizzotto, M., Porter, W., Byrne, D. and Cisneros-Zevallos, L. (2014) Polyphenols of selected peach and plum genotypes reduce cell viability and inhibit proliferation of breast cancer cells while not affecting normal cells. *Food Chemistry* 164, 363–370. DOI: 10.1016/j.foodchem.2014.05.060.

Persimmon

4

Alejandra Salvador[1]*, Cristina Besada[1], and Carlos H. Crisosto[2]

[1]Instituto Valenciano de Investigaciones Agrarias, Valencia, Spain
[2]University of California, Davis, California, USA

Scientific Name, Origin, Current Areas of Production and Importance

Persimmon or kaki, *Diospyros kaki* Thunb., is a tree belonging to the family *Ebenaceae* and native to the Far East. It is also known as Chinese date plum. Other *Diospyros* species, including *Diospyros virginiana* (North America) and *Diospyros lotus*, are used as rootstocks. Persimmon was originally cultivated in the mountains of central China with records of production over 3000 years ago then moved to Japan and Europe (Woolf and Ben-Arie, 2011). For hundreds of years, persimmon growing was especially widespread in Asian countries; nowadays, China, South Korea, Spain, and Japan remain the world's major producers of persimmon (72.3% from China, 6.6% from South Korea, 7.0% Spain, and 4.0% from Japan) (FAOSTAT, 2019). Current production (2018) is 5,430 Mt and increasing, concentrated in Asian countries. In the last few decades, the cultivation of persimmon has expanded significantly, mainly in temperate and subtropical regions all around the world including other Mediterranean countries, and the USA. In these new areas of production, such as Brazil, Italy, Israel, New Zealand, Iran, and Australia, most cultivars came from introductions from Japan. However, new cultivars have arisen as bud sports, developing local cultivars in all of the countries where the crop was established; 'Kaki Tipo' in Italy, 'Rama Forte' in Brazil, 'Triumph' (a cultivar from the USA introduced and grown in Israel) and 'Rojo Brillante' in Spain are cultivars that developed from bud sports and whose fruit are well known as commercial persimmons. Such expansion has occurred in response to export market opportunities for "out-of-season" fruit in traditional markets and "exotics" in non-traditional markets.

*Corresponding author: asalvado@ivia.es

Fruit Composition, Botany and Physiological Characteristics

Persimmon is famous for its nutrition comprising about 80.3% water, 0.6% protein, 0.2% total lipids, 19% total carbohydrates (glucose, fructose, and sucrose), organic acids (citric, malic, and succinic acids), up to 1.5 g total dietary fiber and 7.5 mg ascorbic acid. Persimmon is naturally endowed with bioactive molecules including proanthocyanidins, flavonoids, tannins, phenolics, carotenoids, and dietary fiber (Ito, 1971, 1980; Novillo *et al.*, 2015a). The persimmon fruit (*D. kaki*), which is also known as kaki, is botanically classified as a berry which consists of a rather homogeneous parenchymatous pericarp surrounded by a thin and fragile skin with color that varies from light yellow-orange to different degrees of orange-red making them very attractive. There are many different persimmon cultivars; the shape of the fruit varies from flat to round ('Fuyu' and 'Triumph') to elongated ('Rojo Brillante' and 'Hachiya'), while its weight can be as little as a few grams to more than 0.5 kg depending on the cultivar. Persimmon fruits have been shown to follow a double sigmoid growth curve, consisting of two rapid growth stages, Stage I and Stage III, separated by a period of slow growth (Stage II).

Persimmon cultivars based on sensory perception at harvest are classified into two general groups: (i) non-astringent; and (ii) astringent (sweet persimmons) (Yonemori *et al.*, 2003; Yonemori and Suzuki, 2009; Novillo *et al.*, 2013). Astringency is the sensation that results when tannins bind salivary proteins and cause them to precipitate or aggregate, which leaves a rough "sandpapery" or dry and puckery sensation felt in the mouth. Although fruits of both groups are very astringent when small and immature, non-astringent cultivars lose astringency through their development on the tree. The soluble tannins decrease to levels that are sensorially non-detectable in non-astringent persimmon cultivars prior to when fruit color breaking occurs. The astringency perception remains strong for astringent persimmon cultivars even when fully colored, and only when they are overripe and significantly very soft, do they lose astringency and become edible. Astringent cultivars showed higher soluble polyphenols and greater antioxidant capacity and had higher contents of sugars and organic acids than non-astringent ones (Pesis *et al.*, 1988). Thus, one aspect to consider is that, from the sensory point of view, high levels of astringency due to soluble tannins is an undesirable property when the fruit is consumed fresh. Thus, astringent cultivars must be subjected to postharvest deastringency treatments before being marketed. Otherwise, they must be left on trees until the persimmons are overripe and consequently can be eaten as soft persimmons.

Fruit from astringent cultivars have a high content of soluble tannins even when fully colored (0.5%) and only values close to 0.03% of soluble tannins have been reported in non-astringent 'Jiro' and 'Harbiye' persimmons, while astringent cultivars such as 'Hiratanenashi', 'Rojo Brillante' or "similar types" have shown soluble tannins near 1.0%. A threshold of soluble tannins leading to astringency perception depends on the cultivar, soluble tannins above 0.1% have been found to give an astringent sensation in persimmon cultivars 'Kaki Tipo', 'Lycopersicom', and 'Thiene' and soluble tannins of 0.03% for 'Rojo Brillante' and 'Triumph', making all of these fruit inedible (Salvador *et al.*, 2007; Novillo *et al.*, 2013; Munera *et al.*, 2017). Therefore, it is necessary to apply postharvest treatments to remove astringency prior to the commercialization of those cultivars that are astringent at harvest. An astringency removal treatment based on high carbon dioxide has been commercialized to allow consumers to eat astringent cultivars. This treatment applied to astringent cultivars resulted in a drastic loss of soluble polyphenols and induced changes in total antioxidant capacity, carotenoids, and sugars composition. However, in each group, the fruit astringency of some cultivars is influenced by pollination (pollination variant) and cultivars whose fruits are not affected by pollination (pollination constant). Therefore, persimmon fruits can be classified into four groups:

1. Pollination-constant non-astringent (PCNA) group, which is non-astringent irrespective of seeds being present and fruits can be eaten at harvest, when they are crisp as apples, since their soluble tannins content is low enough for no sensory detection of astringency. Cultivars from this group are the most desirable as they are edible at harvest and no special treatment is required prior to consumption. 'Fuyu' is the main representative of this group, heavily produced and marketed in Japan, Korea, Brazil, and New Zealand (Ikegami *et al.*, 2005) Research efforts are ongoing to develop large-size and attractive PCNA cultivars.
2. Pollination-variant non-astringent (PVNA) group, which is non-astringent at harvest if fruits have seeds, and fruits are not edible when firm if they have not been pollinated. The loss of astringency in PVNA-type cultivars is related to the presence of the seeds that produce acetaldehyde and ethanol, thus causing a coagulation of the tannin, even before the fruit ripens (Sugiura and Tomana, 1983; Yonemori and Matsushima, 1985).
3. Pollination-constant astringent (PCA) group, which is always astringent when firm; the best commercial representative of this group is 'Rojo Brillante' intensively produced in Spain and marketed in Europe. This cultivar became very popular because it produces fruit with large size and bright, attractive red color but needs to be treated to remove astringency before consumption. Other cultivars in this group are 'Hachiya' and "Hachiya types".
4. Pollination-variant astringent (PVA) group, which is also astringent if pollinated, and is non-astringent only around seeds, where there are dark tannin spots. 'Triumph', growing well in Israel and South Africa, and 'Kaki Tipo', grown in Italy, are representatives of this group.

This classification based on pollination variant is important to understand to determine the causes of flesh browning that lead to postharvest losses that are discussed below.

Ethylene production and sensitivity and responses to ethylene

Fruit based on their patterns of respiration and ethylene production during maturation and fruit ripening are classified as climacteric (Salvador *et al.*, 2007). Although persimmons sometimes produce a low ethylene level during the period of maturation and ripening, they are highly sensitive to ethylene, which accelerates softening and thus reduces shelf life and marketability (Itamural *et al.*, 1991; Wills *et al.*, 1998; Kubo *et al.*, 2003; Krammes *et al.*, 2006; Park and Lee, 2008; Besada *et al.*, 2010a).

Non-astringent persimmons produce $<0.1\,\mu l\ kg^{-1}\ h^{-1}$ at 0°C and 0.1–$0.5\,\mu l\ kg^{-1}\ h^{-1}$ at 20°C. 'Rojo Brillante', an astringent cultivar, showed a change in ethylene production during maturation indicating typical climacteric behavior, with maximum ethylene level of $0.04\,\mu l\ kg^{-1}\ h^{-1}$ that coincided with fruit color breaking. However, persimmons are very sensitive to exogenous ethylene exposure that triggers autocatalytic ethylene production ending in softening and ripening. Thus, exposure to $1\,\mu l\ l^{-1}$ and $10\,\mu l\ l^{-1}$ ethylene at 20°C accelerated softening to less than 17.8 N, that limited marketability, after 6 days and 2 days, respectively. For example, in 'Fuyu' even $0.2\,\mu l\ kg^{-1}$ ethylene exposure at 20°C for 1 day affected postharvest life. Lower sensitivity was observed when fruit were stored at 0°C under modified atmosphere packaging (MAP) conditions after ethylene exposure at 20°C (simulating pre-packing exposure for sea freight), where fruit quality was not affected by exposures of 1 day at ≤1 ppm, and only slightly by 2 days at <0.5 ppm (Besada *et al.*, 2010a). As persimmons are climacteric and highly sensitive to ethylene, removing or excluding ethylene during transportation and storage is highly recommended.

Rates of respiration production

In general, harvested persimmons have low respiration rates: 2–4 ml CO_2 $kg^{-1}\ h^{-1}$ at 0°C and 10–12 ml CO_2 $kg^{-1}\ h^{-1}$ at 20°C.

Composition and Health Benefits

Fruit are a good source of carotenoids (β-cryptoxanthin, lutein, violoxanthin, zeaxanthin, and β-carotene), dietary fiber, and vitamins A and C (Ebert and Gross, 1985; Niikawa *et al.*, 2007; Salvador *et al.*, 2007; Del Bubba *et al.*, 2009; Tessmer *et al.*, 2016). Persimmon contains many other bioactive compounds, such as flavonoids, terpenoids, steroids and minerals (Santos-Buelga and Scalbert, 2000; Giordani *et al.*, 2011). Phytonutrients in persimmon include, in addition to the sugars, vitamins and fiber content characteristic of most fruit, large amounts of condensed tannins and polyphenols, which contribute to the high antioxidative potential of these fruit (Ito, 1971, 1980; Zheng and Sugiura, 1990; Senter *et al.*, 1991; Yokozawa *et al.*, 2007; Del Bubba *et al.*, 2009). Genotype influences on health properties have not been addressed; non-astringent cultivars were found to have higher levels of vitamins A and C than astringent cultivars, but the effect of removing astringency on antioxidative activity is not well known. The high contents of tannins, polyphenols, carotenoids, ascorbic acid, and sugars indicate the high potential for health benefits from persimmon consumption. High-molecular-weight-tannins have a greater antioxidative activity than low-molecular-weight tannins. In fact, in some countries like China, persimmon fruits and leaves are traditionally used for many medical purposes such as for treating coughs, hypertension, dyspnea, paralysis, frostbite, burns, and bleeding. Reduction of cardiovascular problems, high cholesterol, diabetes, cancer, and stroke problems have been associated with persimmon consumption. It has been demonstrated that persimmons possess hypolipidemic and antioxidant properties that are attributed to its water-soluble dietary fiber, carotenoids, and polyphenols, with persimmon phenols being mainly responsible for the antioxidant effect of this fruit (Gorinstein *et al.*, 1998, 2011). The persimmon peel has also been shown to be a valuable source of antioxidants that, in diabetic conditions, would reduce the oxidative stress induced by hyperglycemia. The important growth in persimmon production that has taken place in many countries over the last few years, coupled with the higher percentages of fruit destined for exportation, has led to the need to adapt postharvest technology in order to ensure that high quality fruit reach the final consumer.

Quality and Consumer Preferences

Fruit should be free from growth cracks, mechanical injuries, sunburn (Figs 4.1 and 4.2), and decay. High-quality persimmons are medium to large, with uniform skin color from yellow to bright orange. In general, persimmons have subtle taste with very low acidity and weak aroma without a predominant aroma characteristic. Buyers, based on consumer studies, suggest a minimum total soluble solids (TSS) as a prediction of sugars and a potential fruit consumer quality index for non-astringent persimmons. However, it is well known that TSS varies depending on the climate, cultivar, season, region and management, therefore caution should be used with application of TSS as a quality index. The same persimmon cultivars growing in warm production areas will result in higher TSS content ($\geq 20\%$), than fruit produced in cooler environments where fruit may struggle to achieve an average TSS of even 16%. For example, a minimum TSS of 15% is recommended for 'Fuyu' at harvest in New South Wales (Macarthur, 2003),

Fig. 4.1. Onset of sunburn damage symptoms on fruit hanging on the tree. Photo courtesy of Dr. Carlos H. Crisosto.

Fig. 4.2. Advanced sunburn damage. Photo courtesy of Dr. Carlos H. Crisosto.

while 'Fuyu' persimmons can attain TSS ≥18% in Japan and 18–20% TSS in California. A TSS of 21–23% for 'Hachiya' is suggested in California. In general, fruit with good canopy light distribution and/or growing in warm production areas will tend to achieve the best flavor (high TSS and more volatiles). The final fruit flavor perception is not only dependent on sugar (TSS), but also on other sensory components such as volatiles, acidity, texture, etc. which can compensate for a low sugar content. Thus, the application of a strict TSS as a consumer quality index should be done carefully as some persimmons with TSS lower than the requested index level can be highly accepted by consumers, therefore, sales would be lost. What is well known is that TSS is not a reliable quality index for astringent cultivars as tannins have a large contribution to the TSS reading by refractometer adulterating the sugar content in the sample (Salvador *et al.*, 2006). In persimmon, the threshold of soluble tannins leading to astringency perception depends on the cultivar. Soluble tannins at a level of above 0.1% have been found in several persimmon cultivars to confer a strong astringent sensation, and these fruits are not edible. In a sensory study, trained panelists classified persimmons as non-astringent when fruit had a soluble tannins content below 0.03%. This 0.03% astringency perception onset value is in agreement with those reported for both 'Rojo Brillante' and 'Triumph' in which soluble tannin contents higher than 0.4% led to intensely astringent fruit, while values between 0.04% and 0.4% were rated by the panelists as slight to medium astringency (Salvador *et al.*, 2007; Besada *et al.*, 2010a, b, 2014; Munera *et al.*, 2017). In non-astringent cultivars like 'Jiro' and 'Harbiye', values close to 0.03% soluble tannins have been reported at harvest (Taira *et al.*, 1989; Candir *et al.*, 2009). Astringent cultivars also display a gradual reduction of soluble tannins, but it is much less marked than in non-astringent ones. Therefore, the fruits of astringent cultivars have high soluble tannins content even when fully colored and astringency becomes undetectable only in overripened stages, when fruits are completely soft. Astringent cultivars, such as 'Hiratanenashi', 'Rojo Brillante', or 'Tipo', have a soluble tannins content that comes close to 0.5–1% (Taira *et al.*, 1998; Salvador *et al.*, 2007; Del Bubba *et al.*, 2009) during the stage when fruits are completely colored and still firm. Perception of astringency is only lost in overripened stages, when soluble tannins content decrease to values of around 0.03% (Tessmer *et al.*, 2016).

Maturity and Harvest Indexes

In astringent and non-astringent cultivars, the best maturity index is skin color changes (Crisosto *et al.*, 1995; Crisosto, 2004) from green to orange or reddish-orange ('Hachiya'), or green to yellow or yellowish-green ('Fuyu,' 'California Fuyu,' 'Jiro') or green to bright orange-red color for 'Rojo Brillante' (Fig. 4.3). The color changes occurred because

Fig. 4.3. Maturity color changes. Photo courtesy of Dr. Alejandra Salvador.

of chlorophyll degradation and carotenoid biosynthesis (Ebert and Gross, 1985). The principal carotenoids are initially β-cryptoxanthin, zeaxanthin, neoxanthin, lutein, and violaxanthin (Zhou *et al.*, 2011) and the change to red color is due to lycopene accumulation. Sugars accumulate while fruit remain on the tree prior to development of the desired color and becoming too soft. Thus, a long sugar accumulation period will develop persimmons with high flavor for consumers. After color break, most cultivars' skin turns to dark orange-red hues that have been linked to a drastic increase in lycopene content in the most advanced maturity stages (Munera *et al.*, 2017). These skin color changes to achieve the characteristic orange-red color have often been linked to loss of firmness and to a reduction in the soluble tannins responsible for astringency (Tessmer *et al.*, 2016). Gradual fruit softening during ripening is related to microstructural changes in flesh, such as progressive parenchyma degradation with less swollen and more deformed cells. As maturity advances, degradation of cell walls and membranes takes place, and intercellular spaces are filled with solutes, which leads to a generalized loss of intercellular adhesion in the most advanced maturity stages. This reduction in tannins during maturation differs between astringent and non-astringent cultivars. A maximum maturity index based on fruit firmness is used by growers to assist in how late fruit should be hanging prior to harvest. The firmness of the flesh at harvest plays a decisive role in protecting fruit from mechanical damage and preserving the quality of the fruit during postharvest storage and marketing handling. Loss of firmness is an unavoidable fact that occurs eventually; depending on the conditions and time under which the fruit are maintained postharvest. Therefore, the optimum external color of persimmon at harvest must be decided on, not only based on the specific cultivar, but also on the proposed or potential postharvest management and marketing plans. For example, if persimmons are to be marketed for a long-distance market, maturity based on color should be very close to the minimum ideal color rather than an excess of this color and the orchard should use frequent harvests to assure the proper minimum maturity. In most cases, firmness should not be below 21.6 N (measured with an 8 mm tip for 'Fuyu' and similar cultivars). In California, the most used minimum maturity index for 'Hachiya' persimmon is color measured at the blossom-end changing to orange or reddish equal to or darker than Munsell color chart 6.7YR 5.93/12.7 on at least one-third of the fruit's length, with the remaining two-thirds a green color equal to or lighter than Munsell 2.5 GY 5/6. (Crisosto *et al.*, 1995; Crisosto, 2004). For other varieties, fruit must have attained a yellowish-green color equal to or lighter than Munsell 10Y 6/6. In the most popular astringent cultivar 'Rojo Brillante', because of its large size and characteristic bright orange-red color, these changes in color occur with a loss of firmness and a decrease in the soluble tannins responsible for the astringency. To assess the condition of a particular cultivar or a specific orchard, the skin color changes, increase in size and sugars and loss of firmness during maturation can be modeled and used to test the effects of orchard management and chemical treatments on final persimmon quality. Color maturity charts that link 'Fuyu' skin color changes to internal changes are available to assist with determination of the harvesting decision. These skin color and internal changes, including firmness, are location dependent thus a fine-tuning should be carried out to apply this color chart to a particular site.

A chemical and microstructural study comparing the astringency-tannins metabolism process during maturation of two astringent cultivars ('Rojo Brillante' and 'Giombo') with two non-astringent ones ('Fuyu' and 'Hana Fuyu') revealed that the decline in natural astringency during maturation was clearly linked to a tannin insolubilization process inside the tannin cells in both types of cultivars. Light microscopy analysis revealed a much higher soluble tannins content in astringent fruit flesh compared with non-astringent in the early maturity stages and uncovered a tannins insolubilization process in all cultivars during maturation and ripening. In astringent cultivars, the gradual tannins insolubilization process observed led to a progressive decline in soluble tannins with the subsequent astringency reduction. However, in 'Rojo Brillante' only at the last overripe stage (very soft fruit) did tannins insolubilization seem to be completed: precipitated tannins were observed inside the vacuole of tannic cells, no soluble tannins were observed, and no sensorial astringency perception was detected. Nevertheless, in 'Giombo' although substantial tannins insolubilization was also observed, at the over-ripe stage a portion of soluble tannins was still dispersed around tannic cells, which is consistent with the higher soluble tannins value measured and the astringency detected. These fruit at an advanced stage of maturity had levels of soluble tannins of around 0.5%, which have been linked to an intense level of astringency in different sensory studies (Salvador *et al.*, 2007; Besada *et al.*, 2010a, b, 2014; Munera *et al.*, 2017).

Non-astringent cultivars ('Fuyu' and 'Hana Fuyu') displayed precipitated tannins inside tannic cells even from early maturity stages although some soluble tannins were dispersed in parenchymatous tissue. The tannins insolubilization process during maturation was much faster than in astringent cultivars; thus, when fruit is still firm the sensory astringency was not detectable in either cultivar. The study also revealed that both color development and fruit softening during maturation are characteristic for each cultivar although in all of them a strong negative correlation was observed between these two parameters. Fruit softening was associated with a gradual parenchyma structure loss due to a cell wall and membrane degradation process. The maturation-harvest period of persimmon is within 2 or 3 months, and there are benefits to extend the harvest period. An acceleration of fruit maturity has been achieved by preharvest applications of ethephon (Daniell *et al.*, 2002; Kim *et al.*, 2004) and paclobutrazol (PBZ). Delay of maturity was accomplished by applying PBZ or uniconazole (gibberellin biosynthesis inhibitors) and the synthetic cytokinin, N-(2-chloro-4-pyridyl)-N-phenylurea) (CPPU) (Agustí *et al.*, 2003). Also, preharvest applications of gibberellic acid (GA_3) when the fruit is breaking color delayed persimmon maturity 2 weeks later. It must be considered that these kinds of treatments may have negative effects on postharvest life and must be registration approved in some production areas (Lee *et al.*, 1997; Besada *et al.*, 2008a). Thus, it is necessary to conduct specific studies for each cultivar to know the effect of these treatments on fruit maturation and postharvest life. The use of these chemicals should be restricted based on each country's maximum residue limits (MRLs).

Harvesting and Handling

Persimmons are harvested during autumn and proper handling at harvesting (Fig. 4.4) is key to achieve good quality persimmons and determine marketability and profits. The external color is the maturity index commonly used for harvesting persimmons. Fruit needs to be fully developed, firm, and have the cultivar's characteristic color before being harvested. Immature fruit does not soften evenly after harvest and may remain partly astringent and generally lack flavor. Most persimmon

Fig. 4.4. Persimmon harvesting. Photo courtesy of Dr. Carlos H. Crisosto.

cultivars are considered ready for harvest when they display a full orange to orange-red color, with no visible green background (Fig. 4.3).
Depending on the cultivar and the seasonal conditions, the number of
pickings done to complete the harvest varies from one to three (Fig. 4.4).
For astringent and non-astringent cultivars, the best method of harvesting fruit is to cut them from the tree using small clippers, leaving the
calyx attached to the fruit (Fig. 4.5). It is possible to snap fruit from the
tree by hand, but this practice is not highly recommended as it can in-
jure the fruit and the adjoining shoot and result in subsequent decays.
Fruit must be handled very carefully to avoid bruising, which is likely to
become unsightly as fruit ripen, but will also increase decay and cosmetic
marking. Two or three harvests are usually required, depending on fruit
size and color. The harvested fruit should be placed in shallow containers with smooth or padded sides, since persimmons are highly prone
to mechanical damage and blemishes. Special care must be taken with
those cultivars with a pointed apex, such as 'Hachiya', in order to prevent them from causing any damage to other fruit pieces that may end
up with decay caused by *Alternaria* or *Botrytis*.

Packing, Packaging and Grading

After harvesting, astringent and non-astringent persimmons are hauled
(Fig. 4.6) to a packinghouse and gently dumped on a packing line
(Fig. 4.7) where they are cleaned by soft roller brushes. Then, fruit are

Fig. 4.5. Persimmon clipping. Photo courtesy of Dr. Carlos H. Crisosto.

Fig. 4.6. Persimmon hauling. Photo courtesy of Dr. Carlos H. Crisosto.

Fig. 4.7. Persimmon harvesting containers dumping on the packing line. Photo courtesy of Dr. Carlos H. Crisosto.

graded by taking size, shape, firmness, degree of blemishes, overmaturity, immaturity, and color into account. Workers under good lighting and in comfortable working conditions carry out this grading. Grading is usually repeated three times during packing: first, after dumping on the packing line (Fig. 4.8), second after cleaning (Fig. 4.9), and third at the end during placement of the fruit in the containers (Fig. 4.10). Most modern packing lines also include weight or optical sizing (Fig. 4.11). In some operations, shape, firmness, degree of blemishes, and color are optically graded (Fig. 4.12). A waxy cuticle naturally covers the persimmon surface; thus, it is important to keep the roller brushes in good condition to avoid natural wax from being removed (Pérez-Munuera *et al.*, 2009a). Furthermore, persimmons are highly sensitive to mechanical damage, therefore special care must be taken in the design and maintenance of packing line operations. It is advisable to strive to minimize the number and height of drops, equipment transitions should be minimized, and unavoidable impacts should be prevented by cushioning with foam rubber and other materials as persimmon fruit are very sensitive to mechanical impacts. Bruising injuries suffered by persimmon fruit during packing handling have been associated with a faster loss of firmness, weight loss, and decay incidence during cold storage. In addition, the mechanical damage that fruits are exposed to during packing operations has been reported as one of the main causes of persimmon flesh decoloring (Novillo *et al.*, 2014). At the end of the line, the sized fruit are

Fig. 4.8. Persimmon grading starts immediately after the fruit is dumped on the packing line. Photo courtesy of Dr. Carlos H. Crisosto.

Fig. 4.9. Persimmon cleaning using rollers. Photo courtesy of Dr. Carlos H. Crisosto.

Fig. 4.10. Persimmon grading after cleaning. Photo courtesy of Dr. Carlos H. Crisosto.

Fig. 4.11. Persimmon sizing. Photo courtesy of Dr. Carlos H. Crisosto.

packed by hand generally into single-layer trays or two layers according to the current market requirements (Fig. 4.13). Persimmons are packed in one-layer tray packs, two-layer tray packs, 11.3 kg volume fill boxes and clamshells (Fig. 4.14). In general, there are approximately 15–30 persimmons per tray layer and 48–72 persimmons per volume fill box. Containers are palletized into 81 volume fill boxes on one pallet, 120 one-layer-tray-packed boxes on one pallet, and 88 two-layer-tray-packed boxes on one pallet. A desirable size for 'Fuyu' is 230–250 g; 200 g is the minimum marketable size. For astringent persimmons, it is recommended to remove astringency after packing as flesh browning is triggered on fruit after astringency removal going through the packing line.

Fig. 4.12. Persimmon optical grading and sizing. Photo courtesy of Dr. Carlos H. Crisosto.

Mechanical damage symptoms have been studied in the astringent cultivar 'Rojo Brillante'. It has been observed that two different alterations are manifested, depending on the level of astringency of fruits when mechanically impacted: "flesh browning" (Fig. 4.15) and/or "pinkish-bruising" (Fig. 4.16). Flesh browning is manifested as large brown areas of the flesh that extend around the fruit (underneath the skin); browning starts on the surface and then spreads to inner regions. This disorder appears on fruits that have undergone a mechanical impact after astringency was removed by high CO_2 treatment. Pinkish-bruising is expressed as isolated areas of pulp (close to the skin) in which the habitual orange color has turned pinkish, and is detected in fruits that have suffered mechanical damage when astringent (Novillo *et al.*, 2014, 2015c). Oxidative stress has been reported as the key mechanism that lies behind these alterations (Novillo *et al.*,

Fig. 4.13. Persimmon packing in single-layer trays. Photo courtesy of Dr. Carlos H. Crisosto.

Fig. 4.14. Persimmon packing in clamshells. Photo courtesy of Dr. Carlos H. Crisosto.

2015b). Not only mechanical damage itself but also CO_2 deastringency treatment results in reactive oxygen species (ROS) accumulation in the damaged flesh of fruit. Under such oxidative conditions, tannins, which are initially uncolored, undergo an oxidation process and become colored. When the insoluble tannins of fruits that are submitted

Fig. 4.15. Persimmon flesh browning caused by mechanical damage during harvest. Photo courtesy of Drs. Salvador and Besada.

to deastringency treatment are oxidized, they acquire a brown coloring (flesh browning), while the oxidation of soluble tannins present in astringent fruits allows them to acquire a pinkish color ("pinkish-bruising"). This pattern of response to mechanical damage is shared by most astringent cultivars. Although the intensity of these disorders depends on the cultivar, they all seem to manifest browning or pinkish-bruising, depending on the level of astringency at the time of the mechanical impact. Non-astringent cultivars, however, show less sensitivity to manifesting such alterations associated with mechanical damage (Novillo *et al.*, 2015c). As fruit is highly sensitive to manifesting such alterations, proper conditions on packing lines to avoid impacts will assure high quality fruit. From a commercial point of view, flesh-browning manifestation implies greater loss of quality than pinkish-bruising, as the former spreads throughout the fruits, while the pinkish disorder is manifested only on localized areas that have suffered strong impacts. Therefore, in order to reduce fruit sensitivity to browning, applying a deastringency treatment is recommended after carrying out packing operations. Persimmons will be submitted to special postharvest handling such as astringency removal in the case of astringent cultivars, and/or 1-methylcyclopropene (1-MCP) depending on their market destination (Novillo *et al.*, 2014).

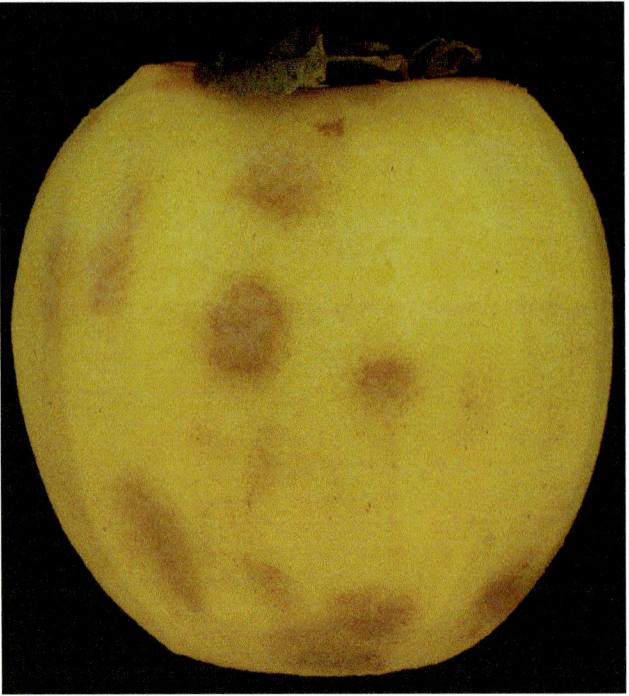

Fig. 4.16. Persimmon pinkish bruising due to mechanical damage during harvesting. Photo courtesy of Drs. Salvador and Besada.

Persimmon Disorders

Calyx separation

In some cultivars and growing conditions, calyx separation or 'calyx cavity' develops (Glucina, 1987). This disorder occurs late in the season, during the final phase of the double sigmoidal growth curve. It is thought that the fruit expands faster than the calyx and this results in the flesh separating from the calyx tissue resulting in potentially large gaps (up to 5 mm or more), which may even encircle the whole calyx. This leads to more rapid softening of non-stored persimmons and increased levels of chilling injury (CI) in stored fruit. Thus, commercial recommendations are to remove fruit with this disorder during packing although detection of all but the most severe disorders can be difficult. These cavities also provide a refuge for insects such as caterpillars, mealybugs and even earwigs, ending with potential *Alternaria* decay. The problem is more common in large 'Fuyu' fruit because of the greater fruit expansion. It is considered to occur more on young trees, and generally declines after about 15 years. Excessive nitrogen fertilization should be avoided and thinning trees early in the season will enhance calyx growth and help prevent this disorder.

Flesh browning

Flesh browning is one of the main physiological disorders that causes important quality losses during the postharvest life of persimmons. Different types of flesh browning have been described at harvest and during postharvest handling. The expression of this flesh-browning disorder can be associated with different causes. In some cases, the appearance of the browning disorder can give us clues to the cause. In other cases, different factors may lead to visually similar browning disorders (Besada *et al.*, 2018).

Natural flesh browning (at harvest)

There are two types of natural flesh browning at harvest linked to the presence of seeds in PVA-type and PVNA-type cultivars and a third type of flesh browning manifested in unseeded fruits from some PVNA cultivars (Fig. 4.17). For the first two cases, the flesh browning at harvest in PVA-type and PVNA-type cultivars appears to be a consequence of acetaldehyde production by seeds; acetaldehyde causes condensation or coagulation of soluble tannins that become insoluble and oxidized, visible as flesh browning. In the PVA-type cultivars, in which the production of acetaldehyde is limited, tannins insolubulization and the consequent flesh browning only happens in the flesh area surrounding the seeds. For the third case, frequently occuring in some unseeded PVNA cultivars, such as 'Fuju', brown specks are visible in the flesh. This flesh-browning damage has been associated with high temperature ($\geq 30°C$) during the last stages of fruit development and maturation (Mowat and George, 1996).

Postharvest flesh-browning disorders

FLESH BROWNING TRIGGERED BY MECHANICAL DAMAGE. Mechanical damage to persimmons during postharvest handling is one of the known factors involved in flesh browning. Persimmons are generally very sensitive to mechanical damage, especially during packing when fruit may exhibit brown or pinkish areas around the flesh. The mechanism of this alteration has been studied, and the oxidation of tannins has been reported as a key process. Further details on flesh browning due to bruises during harvesting (Fig. 4.15) and pinkish-bruising (Fig. 4.16) are described in the "Packing, Packaging and Grading" section above.

FLESH BROWNING DUE TO CO_2 OVEREXPOSURE. The exposure of persimmons to the CO_2 desastringency treatment for excessively long periods can also lead to flesh browning, even in the absence of mechanical damage. This type of browning becomes visible around the core of the fruit, mainly in the calyx end of the fruit (Fig. 4.18). The intensity of this browning increases when persimmons are exposed to low temperature after deastringency treatment. Maturity also is an important factor that affects this flesh

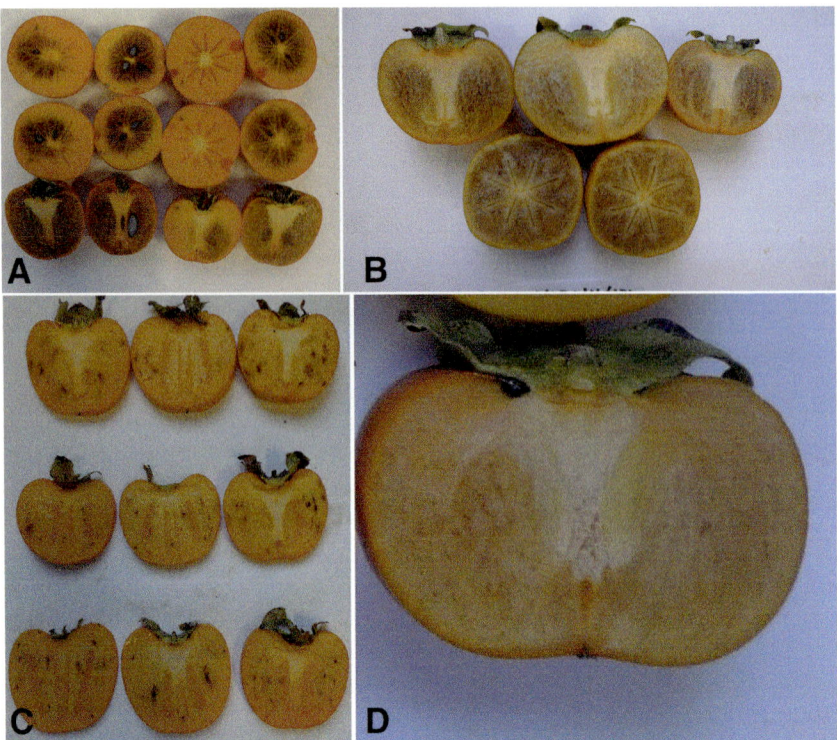

Fig. 4.17. Different types of natural flesh browning. (**A**) Flesh browning associated with the presence of seeds in PVA cultivars. (**B**) Flesh browning associated with the presence of seeds in PVNA cultivars (**C**) Flesh browning manifested in unseeded fruits from some PVNA cultivars. (**D**) Flesh browning detected in some PCNA cultivars. Photos courtesy of Drs. Salvador and Besada.

browning incidence and intensity since a longer CO_2 treatment period is required for advanced maturity persimmons.

FLESH BROWNING DUE TO STORAGE UNDER LOW O_2 AND/OR HIGH CO_2 ATMOSPHERES.   Another factor frequently associated with flesh browning is storage under controlled or modified atmospheres; in this case, the flesh browning occurs mainly in the flesh of the fruit. The optimum atmosphere conditions to prolong storage will depend on the cultivar; thus, the same atmosphere can be effective for one cultivar and lead to disorders in another one. Flesh browning is one of the main disorders associated with persimmon storage at low O_2 levels (Burmeister *et al.*, 1997; Park and Lee, 2008). The tolerance to reduced oxygen atmospheres depends on the cultivar. Thus, for example, in a study in which 'Triumph' and 'Rojo Brillante' persimmons were submitted to the same controlled atmosphere (CA) storage conditions (4–5% CO_2+ N_2), internal browning was manifested in 'Rojo

Fig. 4.18. Persimmon flesh browning due to long exposures to CO_2. Photo courtesy of Drs. Salvador and Besada.

Brillante' (Fig. 4.19) but not in 'Triumph'. In the case of 'Rojo Brillante', the disorder appeared after 75 days of storage irrespective of 1-MCP treatment, and symptom severity increased when persimmons were transferred to shelf-life conditions (Besada *et al.*, 2014). Different flesh-browning symptoms have been documented as consequences of low O_2 exposure. For example, "stylar" end browning was observed in 'Fuyu' stored under a modified atmosphere (0.1–0.3% O_2; Park and Lee, 2008). Finally, another factor associated with flesh browning is high or low temperature exposure. The application of heat treatments such as hot water or hot air can cause damage and discoloration in the fruit.

Low-temperature-storage chilling injury (CI)

CI is a term used to describe the physiological damage that occurs in many plants and fruits, particularly those of tropical and subtropical origin, as is the case of persimmon fruit, because of their exposure to low but non-freezing temperatures. As mentioned, the sensitivity of persimmon fruit to CI is cultivar and temperature-time dependent; commercially important cultivars such as 'Fuyu', 'Sugura' or 'Rojo Brillante' are highly chilling sensitive, whereas other cultivars such as 'Triumph' or 'Hachiya' are less

Fig. 4.19. Persimmon low oxygen damage. Photo courtesy of Drs. Salvador and Besada.

sensitive (MacRae, 1987; MacRae, 1987; Collins and Tisdell, 1995; Arnal and Del Río, 2004). 'Fuyu' and similar non-astringent types are sensitive at temperatures between approximately 5°C and 15°C; and they will exhibit flesh browning and softening. Thus, CI can be a major cause of deterioration of 'Fuyu' persimmons during marketing (Fig. 4.20). Symptom development is greatest at 5–7°C and still occurs when held between 2°C and 15°C (Crisosto, 2004). CI symptoms, as well as their incidence and severity, depend on the cultivar, storage temperature, and duration. A higher CI incidence has been reported for cultivars 'Rojo Brillante' (Salvador *et al.*, 2005, 2006) and 'Fuyu' (Krammes *et al.*, 2006) when fruits were picked at early maturity stages. In general, CI symptoms became more severe after transferring fruits from low to ambient temperatures, although they can also be exhibited during cold storage. On transfer to warm temperatures, symptom severity (flesh softening, browning, and water-soaked appearance) increases and renders fruit unmarketable (Woolf *et al.*, 1997a; Yonemori and Suzuki, 2009; Zhang *et al.*, 2010). However, in the case of short cold storage periods, invisible CI symptoms during cold storage may become visible after transferring fruit to warm temperatures. Only during prolonged cold storage periods can CI symptoms eventually appear prior to warm up. Expression of CI symptoms depends on cultivars and in most cases is related to changes in texture. Thus, the first CI symptom in 'Rojo

Fig. 4.20. Persimmon chilling injury (CI) on cultivar 'Fuyu'. Photo courtesy of Drs. Salvador and Besada.

Brillante' and 'Suruga' is a dramatic loss of firmness (Collins and Tisdell, 1995; Arnal and Del Río, 2004). This loss of firmness is not visible while persimmons are stored at cold temperatures, however, a fast softening occurs upon transfer to warm temperatures. Furthermore, in 'Rojo Brillante' after a storage period at 1°C, cell wall changes caused by low temperature induced a hard and gummy texture prior to loss of firmness (Salvador *et al.*, 2005, 2008). Other symptoms associated with CI are flesh browning or flesh darkening, starting in the center of the fruit flesh and diffusing outward (Fig. 4.21). In fruit harvested in the Valencia production area, formation of nodules as a result of hard tissue areas in the stem area were observed (Arnal *et al.*, 2008). In 'Fuyu' or 'Suruga' persimmons harvested in New Zealand flesh gelation (Fig. 4.20), dark flesh, and skin transparency have been reported as CI symptoms (MacRae, 1987; Collins and Tisdell, 1995; Woolf *et al.*, 1997a). Other symptoms that have been associated with development of CI are losses in fruit flavor, sweetness, juiciness, and titratable acidity or color mottling and skin translucency.

In anatomical-biochemical studies, CI manifestation has been related to changes in the cellular structure as accelerated cell wall solubilization of chilling-injured fruit has been reported in 'Fuyu' (Grant *et al.*, 1992). A microstructural study in 'Rojo Brillante' showed that the drastic flesh softening, as a CI symptom, was associated with cell wall material degradation and

Fig. 4.21. Persimmon CI symptoms on cultivar 'Rojo Brillante'. Photo courtesy of Drs. Salvador and Besada.

the loss of intercellular adhesion (Pérez-Munuera *et al.*, 2009b). Likewise, it was reported that the primary cell wall and the middle lamella could not normally be dissolved in chilling-injured fruit when transferred to normal temperatures after cold storage (Luo and Xi, 2005). Recently, it has been documented that the low-temperature storage of persimmon leads to gradual oxidative stress, as well as major hydrogen peroxide (H_2O_2) accumulation, and sharp increases in catalase, peroxidase, and lipoxygenase (LOX) activities that were linked to the manifestation of CI symptoms (Khademi *et al.*, 2014; Novillo *et al.*, 2015b). Many studies have focused on finding solutions to control CI in sensitive persimmon cultivars. Some tested postharvest strategies to alleviate CI symptoms and to allow cold storage to be prolonged in persimmon are discussed below.

Temperature Management

CI in 'Fuyu' persimmons is controlled by avoiding exposure of fruit to temperatures between 2°C and 15°C. However, the standard recommended storage temperature to delay CI and reduce softening is at least 0°C, and some countries recommend even lower temperatures such as −1°C (Woolf and Ben-Arie, 2011). Higher temperatures (3–8°C) lead to increased CI in

'Fuyu'. Since maintaining temperatures around or below 0°C is commercially important, determining the freezing point is required to avoid freezing damage. Determination of freezing points varied from −3.2°C to −1.3°C with a median of −2.0°C, the variation being probably attributable to TSS level in the fruit. Exposure to ethylene >0.2 µl l⁻¹ during postharvest handling should also be avoided. The general agreement is that the maximum storage life of persimmons is generally up to 12–16 weeks and it varies according to cultivars, growing regions, and management. Under air storage, ethylene-free, the non-CI sensitive 'Triumph' has 8 weeks postharvest life at −1°C. 'Fuyu' developed CI within 2 weeks below 8°C, and 'Rojo Brillante' stored well at 15°C for 3–4 weeks. For long-term cold storage, the field heat in persimmons should be removed as soon as possible after harvesting by a forced-air cooling system. Proper application of new technologies such as optimizing temperature management, 1-MCP, and/or modified atmosphere storage to overcome the current main storage limitations such as softening and/or CI problems are extending postharvest life. Postharvest life under optimum temperature and relative humidity (RH) in ethylene-free air is 3 months, whereas fruit can be stored for up to 5 months using ethylene-free CA (3–5% O_2 and 5–8% CO_2). Storage of the fruit is a common practice to supply markets according to demand. Depending on the storage period, storage temperature and cultivar, different postharvest technologies are routinely applied in order to preserve the fruit quality. Furthermore, persimmon shipping to export markets can take a long time that implies the need to apply new technology and maintain optimum transportation conditions in order to assure persimmon quality at arrival.

Special Storage Treatments

Ethylene removal

As persimmon rate of softening and CI sensitivity increase with exposure to ethylene, several methods to eliminate ethylene action are used. Ethylene avoidance, ethylene removal by constant fresh air exchanges, ethylene destruction using ozone, ultraviolet (UV) light and others. Among them, the most effective is using an ethylene sensitivity blocker chemical named 1-MCP.

1-Methylcyclopropene (1-MCP) application

Several studies have reported a positive effect of the postharvest application of 1-MCP on reducing CI symptoms in different persimmon cultivars. The recent availability of the inhibitor of ethylene perception, 1-MCP, which is a non-toxic (but not registered as an organic product) antagonist of the ethylene hormone that binds to the ethylene receptor after treatment, has become an important tool for persimmon postharvest. 1-MCP applied prior to cold storage has been demonstrated to alleviate softening and gelation, which are the main symptoms in cultivars sensitive to low temperatures (Salvador et al., 2004; Kim and Lee, 2005; Yonemori and Suzuki, 2009). This effect of 1-MCP has been associated with preserving both the integrity of cell walls and adhesion between adjacent cells (Pérez-Munuera et al., 2009b) and reduces membrane permeability (Yonemori and Suzuki, 2009) not only throughout cold storage, but also when fruits are transferred to shelf-life temperatures. Reducing CI symptoms in persimmon achieved by 1-MCP treatment has also been attributed to modulation of ROS scavenging enzymes. In 1-MCP-treated persimmons of the cultivar 'Rojo Brillante', alleviation of CI symptoms was linked to lower peroxidase activity levels and also to enhanced catalase enzyme activity, which resulted in slower H_2O_2 accumulation (Khademi et al., 2014; Novillo et al., 2015b). Similarly, Yonemori and Suzuki, 2009 have reported in 'Fuyu' that the application of 1-MCP maintained greater antioxidant enzymes activity, such as superoxide dismutase (SOD) and catalase, and also contributed to lowering the activity of oxidative enzymes, such as polyphenol oxidase (PPO) and peroxidase. Nowadays, 1-MCP treatment is routinely applied in the industry when fruits are cold stored. Moreover, the combined use of 1-MCP and modified atmosphere storage has been shown to prolong the retention of firmness and to reduce CI in 'Fuyu' persimmon (Kim and Lee, 2005; Argenta et al., 2009). The combination of GA_3 preharvest treatment with postharvest 1-MCP application allowed 'Rojo Brillante' persimmon to be stored for longer than when each treatment was singly applied and thus, delayed CI symptoms injury and extended storage time (Besada et al., 2008a). The exposure of persimmons to low ethylene levels triggers softening and color evolution, and aggravates CI during storage. Upon removal from cold storage, persimmon ethylene production and respiration rate increase; and CI symptoms are highly visible. It has been reported that even a very low ethylene concentration induces CI symptoms and softening. In most astringent and non-astringent persimmon cultivars, application of 1-MCP at a concentration of 100–1000 nl l⁻¹ (ppb) delays softening and color evolution of persimmon fruit during its shelf life (Besada et al., 2014). The beneficial effect of 1-MCP extending postharvest life is cultivar and maturity dependent. 1-MCP application has been successfully tested on 'Tonewase', 'Saijo', 'Hiratanenashi', 'Rojo Brillante', 'Triumph', 'Nathanzy', 'Rendaiji', 'Matsumotowase-Fuyu', 'Bianhua', 'Qiandaowuhe' and 'Fuyu' astringent and non-astringent cultivars. In most cases, the postharvest life period can at least be doubled by applying the optimum concentration of 1-MCP. To attain the best 1-MCP results, persimmon should be treated as soon as possible after harvesting; up to 12 h after harvesting might be acceptable in commercial practice. In the case of astringent cultivars, 1-MCP may be applied during the treatment to remove astringency blended with high CO_2 concentrations. In 'Rojo Brillante' persimmon, 500 nl l⁻¹ of 1-MCP applied prior to storage alleviated CI and softening, allowing storage for 30 days, while untreated fruit showed loss of commercial firmness after 10 days under the same storage conditions. A combination approach of GA_3 preharvest treatment with postharvest 1-MCP application allowed 'Rojo Brillante' persimmon to be stored for longer than when each treatment was applied alone, delaying the symptoms of CI, softening, and extending the storage time for up to nearly 3 months (Besada et al., 2008b).

Controlled atmosphere (CA) and modified atmosphere packaging (MAP)

CA and MAP are postharvest technologies that modify O_2, CO_2 and/or C_2H_4 concentrations in the atmosphere surrounding the commodity to maintain quality and extend the commercial life of fresh fruits and vegetables. CA is generally applied in specialized sealed storage chambers that allow the atmospheric composition to be constantly controlled. MAP is the practice of modifying the composition of the internal atmosphere of a package, which can be carried out by active or passive modification. In the case of active modification, the target atmosphere is established by creating a slight vacuum and replacing the atmosphere inside the package with the desired gas mixture. Additionally, absorbers may be included in the package that scavenge O_2, CO_2 or ethylene in order to control the concentrations of these gases. A passive modification of the atmosphere is attained through the respiration of the commodity within the package and depends on the characteristics of both the commodity and the packaging material (Fig. 4.22). The proper use of MAP and CA may supplement proper temperature management to delay senescence, reduce sensitivity to ethylene, alleviate physiological disorders, such as CI (Besada et al., 2015), or control insects and decay incidence. Successful CA and MAP in persimmons must maintain O_2 and CO_2 at near optimum levels to attain the beneficial effects of a modified atmosphere without exceeding the limits of tolerance, which may increase the risk of physiological disorders, and other detrimental effects. Using MAP or CA should be considered to complement storage at proper temperatures.

In persimmon fruits, most research has focused on using modified atmosphere packages, inside which the desired atmosphere is generated passively during cold storage of fruit (Fig. 4.22). Satisfactory results have been obtained with polyethylene or low-density polyethylene bags in 'Fuyu' and 'Rama Forte' (Brackmann et al., 1997, 2006). One of the main factors that limits longer storage life under MAP conditions is the accumulation of ethanol and acetaldehyde, which causes off-flavors to develop and may also result in tissue browning (Ben-Arie et al., 1991).

The effect of CA on extending storage has been widely studied in some cultivars, like 'Fuyu' (Ben-Arie et al., 1997). Although some atmospheres have resulted in reducing CI, they may lead to fruits that manifest external or internal browning (Burmeister et al., 1997; Brackmann et al., 2006).

Fig. 4.22. Use of modified atmosphere packaging (MAP) on persimmons. Photo courtesy of Dr. Carlos H. Crisosto.

The incidence of skin and flesh disorders is the main limitation to storing 'Fuyu' in CA, and such disorders are reported to be due mainly to low O_2 levels and not to high CO_2 levels (Park and Lee, 2008; Woolf and Ben-Arie, 2011). A recent study in 'Fuyu' has reported that the short-term high CO_2 treatments based on fruit exposure to high CO_2 concentrations relieve CI symptoms by preserving the integrity of cell walls and the plasmalemma (Besada *et al.*, 2015). In the last few years, much interest has been shown in introducing the use of CA to some cultivars like 'Rojo Brillante'. One example is an atmosphere based on 97% N_2 +3% air, which led 'Rojo Brillante' fruits to lose astringency and allowed fruit conservation for 30 days at 14°C (Arnal *et al.*, 2008). The use of an ultralow oxygen (ULO) atmosphere (1.3–1.8% O_2) removed astringency in 'Rojo Brillante' when applied at 14.5°C but did not control CI at 1°C or 10°C (Orihuel-Iranzo *et al.*, 2010). Other CAs based on 4–5% O_2+ N_2 offered no additional benefits for retarding onset of CI in 'Rojo Brillante' but prolonged the storage of 'Triumph' by alleviating fruit softening and flesh gelling (Besada *et al.*, 2014). Other authors have reported that CA (1–1.5% O_2 and 1.5–3% CO_2) storage offers the benefit of delaying softening and retarding decay development in 'Triumph'; nevertheless, fruit shelf life became inversely proportional to the storage period length (Tsviling *et al.*, 2003).

The use of passive MAP bags is a common commercial practice in Korea, Japan, and New Zealand. Satisfactory results have been obtained with polyethylene or low-density polyethylene bags (20–80 μm) for 'Fuyu' and 'Rama Forte'. Thus, in several countries, such as New Zealand, Korea, and Japan, 'Fuyu' persimmons are routinely stored in MAP by sealing the fruit in a 60 µm polyethylene bag that delays and reduces the CI symptoms allowing prolonged cold storage. However, exposure of MAP packages to high temperature can increase ethanol and acetaldehyde levels and will override the benefits of MAP by causing development of off-flavors and tissue browning. Low O_2 of 3–5% delays softening; CO_2 at 5–8% helps retain firmness and can reduce CI symptoms on non-astringent 'Fuyu' and other non-astringent CI-sensitive cultivars. Exposure to O_2 levels <3% during storage for longer than 1 month can result in failure of fruit to ripen and development of off-flavors. Tolerance to high CO_2 and low O_2 levels varies among cultivars and growing areas. In California, exposure to CO_2 >10% during 'Fuyu' storage for longer than 1 month can cause flesh-browning discoloration and lead to development of off-flavors. It has been reported in New Zealand, that excessive CO_2 exposure can induce skin browning or onset of flesh browning starting at the center of the fruit. A very careful study carried out in Israel using their popular astringent persimmon 'Triumph' confirmed the commercial benefits of delaying fruit softening including: (i) retarding decay development (caused by *Alternaria*); (ii) extending postharvest life beyond 3 months; and (iii) the extra benefit of removing astringency. The CA levels required to achieve these advantages were 1.0–1.5% O_2 and 1.5–3.0% CO_2. In this cultivar ('Triumph') grown under Israel conditions and CA with O_2 below 1.0%, ethanol and acetaldehyde accumulation were excessive and off-flavors developed after 3 months at −1°C. CO_2 at 3% or 5% controlled *Alternaria* incidence, but unfortunately, the CO_2 exposure ≥3% induced flesh browning, and at this level the rate of softening was not well controlled, thus, limiting postharvest life and marketing of 'Triumph' to no longer than 3 months. The use of CA to avoid CI was also tested in a CI-susceptible astringent cultivar in Spain. By controlling softening and avoiding CI, the postharvest life of 'Rojo Brillante' was extended to 30 days at 15°C under low oxygen (3% O_2, 97% N_2). According to the information provided above, optimum CA conditions are dependent on the cultivar, the physiological state, the atmospheric composition, the storage temperature and the storage time, although some general guidelines have been drawn up in the case of persimmon fruit, which point to the range of 3–5% O_2 and 5–8% CO_2 as being the appropriate atmospheric composition to maintain quality and extend cold storage.

Heat pretreatments

Other treatments reported to be effective to alleviate CI in persimmon fruits are hot water or hot air applications prior to cold storage (Lay-Yee *et al.*, 1997; Woolf *et al.*, 1997b; Besada *et al.*, 2008b). The response of fruit to heat treatments strongly depends on the cultivar. Before cold-storing 'Fuyu', hot air and hot water treatments (HWTs) reduced the flesh gelling and flesh softening associated with CI (Burmeister *et al.*, 1997; Lay-Yee *et al.*, 1997; Woolf *et al.*, 1997a, b). However, heat damage (mainly external and internal browning) was associated with application of heat treatments (Woolf *et al.*, 1997b). In other cultivars, such as 'Rojo Brillante', the effectiveness of HWTs has been reported to depend on the maturity stage of fruits at harvest (Besada *et al.*, 2008b). HWTs applied to fruits at an early maturity stage reduced CI and preserved fruit firmness and quality. However, when these treatments were applied to fruits at more advanced maturity stages, they caused irreversible epidermal breakage and external browning. Tolerance to chilling temperatures that HWTs confer to persimmon fruits has been associated with relevant changes in cell wall degrading and antioxidant system enzymes. In 'Rojo Brillante', HWTs have been reported to downregulate pectin methylesterase and polygalacturonase activities, which results in greater cell wall integrity and, therefore, alleviation of the fruit softening symptom. Moreover, the changes observed in peroxidase and catalase enzymes suggest that HWTs confer greater reactive oxygen scavenging capacity to fruits and may also be implicated in alleviating CI symptoms (Khademi *et al.*, 2014).

Treatments to Remove Astringency (Desastringency)

Astringency is defined as "the complex of sensations due to the shrinking, drawing or puckering of the epithelium tissue covering the tongue as a result of its exposure to tannin acids". The sensation of astringency produced by astringent persimmons is due to their soluble tannins content and the intensity of this perception depends on the concentration of soluble tannins in the flesh of the fruit. In non-astringent cultivars, the content of soluble tannins drops to undetectable levels during maturation, while it remains high in astringent-type cultivars. Soluble tannins of persimmon fruit are accumulated in the vacuoles of so-called "tannin cells" and when astringent-type persimmons are eaten, the soluble tannins are released in the mouth which leads to the formation of insoluble complexes of tannins due to the reaction with salivary proteins (Yonemori *et al.*, 2003). Protein-soluble tannin complexes lead to a decrease in salivary lubrication between various oral surfaces, resulting in the sensation of astringency. Thus, for fresh consumption, astringent persimmon cultivars must be submitted to astringency removal postharvest treatments before being marketed.

The postharvest application of deastringency treatments is a necessary requirement to commercialize fruits from astringent cultivars. As previously discussed, natural astringency loss occurs in overripened fruit. Therefore, one of the traditional methods to remove astringency at harvest consists of treating fruit with ethylene (10 ppm at 20°C) or ethephon water dipping (50–500 ppm) to enhance the tannins complex formation and ripening process. Unfortunately, astringency removal occurs in parallel with a drastic loss of firmness; therefore, fruits with astringency removed are commercialized when very soft. These soft fruits have a very limited postharvest life with handling limitations, and are usually marketed to local markets, however, in some areas of the world, there is a niche market for these types of fruit. Because of the ethylene treatment limitations, other treatments to remove astringency that do not trigger flesh softening such as exposing fruits to alcohol, carbon dioxide (CO_2), nitrogen, or warm water were developed and established for astringency to be removed (desastringency treatment). It is claimed that the biological basis of these anaerobic treatments is to create flesh anaerobic conditions and accumulate acetaldehyde. The high acetaldehyde levels will polymerize soluble tannins (responsible for astringency) to form insoluble compounds, which are perceived as non-astringent (Matsuo and Ito, 1982; Sugiura and Tomana, 1983; Pesis *et al.*, 1988; Taira *et al.*, 1989, 1997, 1998). An anatomical work described the process of tannins insolubilization-polymerization induced by acetaldehyde production at the microstructural level. During desastringency treatment, insoluble material appeared inside the vacuoles of some tannic cells, which were initially filled with soluble material (Salvador *et al.*, 2007; Novillo *et al.*, 2015a) corroborating the hypothesis that acetaldehyde is polymerizing the soluble tannins.

Among the treatments based on exposing fruits to anaerobic conditions, carbon dioxide has been commercially adopted for its effectiveness. Astringency removal by alcohol has a limited used, but it is important for some ethnic groups. The ethanol application was developed and adapted in Japan because this treatment yields dark flesh color, high prices are obtained as this is an indication of high labor costs that are involved in bagging the fruit and applying the ethanol treatment. Most research that has compared the effectiveness of ethanol and CO_2 methods revealed that CO_2 treatment is significantly more effective in removing astringency than applying ethanol (Taira *et al.*, 1992a, b). This is because of the fast acetaldehyde accumulation in CO_2-treated fruit compared with fruit treated with ethanol (Itamural *et al.*, 1991; Taira *et al.*, 1992a, b; Tanaka *et al.*, 1994; Yamada *et al.*, 2002). It has been reported the combined use of high levels of CO_2 and ethanol treatments may increase the efficiency of the deastringency process (Taira *et al.*, 1992a). This combined treatment can be a good alternative in those cases in which CO_2 treatment has to be prolonged to ensure complete astringency removal when treating fruits at an advanced maturity stage and/or when the process is applied at temperatures below 20°C. Currently, the CO_2 removal treatment is widely used commercially for 'Rojo Brillante' in the Spanish industry. For the application of carbon dioxide, a specialized chamber (Fig. 4.23) with a carbon dioxide detector, efficient airflow and venting is used (Fig. 4.24). CO_2- or N_2-enriched atmospheres successfully remove astringency from several persimmon cultivars. In both N_2 and CO_2 treatments, the higher the concentration and the longer the exposure time, the faster the rate of astringency decline. High

Fig. 4.23. Carbon dioxide and 1-MCP application chamber. Photo courtesy of Dr. Carlos H. Crisosto.

Fig. 4.24. Chamber with efficient air flow and venting that contains temperature and relative humidity (RH) controls, a heater, and a carbon dioxide application and detector system. Photo courtesy of Dr. Carlos H. Crisosto.

N_2 treatment is most effective for some cultivars and high CO_2 for others. In all cultivars, the level of soluble tannins decreases as acetaldehyde increases. However, longer exposure time triggers firmness loss. The success of the carbon dioxide treatment depends on temperature, CO_2 concentration, and treatment duration. Several studies have dealt with optimizing these parameters according to the cultivar and other factors. An efficient commercial method widely used to remove astringency while maintaining a high degree of fruit firmness involves holding the fruit in airtight chambers under 80–95% CO_2 at 20°C (fruit temperature) and 90% RH for 1–2 days. With major varieties like 'Rojo Brillante' and 'Triumph', the standard duration is 24 h. Depending on factors such as temperature, CO_2 concentration, and the fruit maturity stage (Taira *et al.*, 1990), this duration could be shorter or longer. The optimum treatment is when the minimum duration is used to ensure that astringency-free fruits reach consumers. It is important to report that exposing fruits to CO_2 deastringency treatment for excessively long periods may result in internal flesh browning. This browning becomes visible around the core of fruits, mainly in the area that surrounds the calyx (Fig. 4.18). It has been observed that the intensity of this disorder associated with CO_2 application increases if fruits are stored at low temperature after removing astringency. In order to avoid this browning, fruits were submitted to the CO_2 treatment for the number of hours required to remove astringency. Therefore, planning the treatment application so that fruits in a CO_2 chamber are at a similar physiological stage is highly recommended because, in this way, CO_2 overexposure is avoided, as all the fruits in the chamber require similar treatment duration. The correct application of these astringency removal postharvest techniques has been crucial in the increase in the number of persimmon plantations in those countries where astringent cultivars are grown.

The astringency removal process consists of two phases that are temperature dependent. The first phase (induction) occurred on fruit under high CO_2 concentration for a given period when a series of required reactions occurred. In the second phase, the presence of CO_2 is not essential when astringency gradually disappears (Gazit and Adato, 1972; Matsuo and Ito, 1977). The optimum temperature to apply the induction treatment has been widely investigated. However, limited information is available on the effect of temperature on phase two. Fruit maturity also plays an important role in the success of the treatment (Taira *et al.*, 1990; Novillo *et al.*, 2013). In most cases, it is more difficult to remove astringency in fruit at advanced maturity than in mature fruit. In these persimmons to remove astringency a long treatment is needed that can trigger softening of the fruit (Besada *et al.*, 2010b). The cold storage period also affects the success of treatment, astringency is quick and easy to remove on freshly harvested fruit, while a longer treatment is required for fruit that have been in cold storage at 15°C for a long period (Besada *et al.*, 2010b; Novillo *et al.*, 2013). This lack of treatment response is explained by the structural changes at the cell level occurring in cold storage. As cold storage is prolonged, degradation of the initial cell structure of the flesh occurs and intercellular spaces, which were initially empty, are progressively filled with cell material. It has been hypothesized that CO_2 diffusion through intercellular spaces becomes more difficult and leads to a low anaerobic respiration rate and, consequently, to less acetaldehyde accumulation. Therefore, it results in incomplete tannins insolubilization, with the subsequent residual astringency sensation (Salvador *et al.*, 2008). Preharvest stresses have also been observed to affect the effectiveness of deastringency treatments. For example, fruit from trees exposed to extreme water stress or to intense salinity conditions do not properly respond to deastringency treatment (Nakano *et al.*, 1997; Besada *et al.*, 2016).

Fruit Pathological Problems

One of the most important postharvest diseases in persimmon is *Alternaria* black spot (ABS) disease caused by *Alternaria alternata*. Fruit affected by this disease show black spots — firm, dry stains of varying sizes and shapes — found below the calyx or on any point of the fruit skin surface (Fig. 4.25). Under field conditions, ABS develops saprophytically on dead organic matter, leaves, shoots, and whole plants. On persimmon fruit that developed small skin cracks due to excessive rain, high humidity, and/or mealybug injury, infection can remain beneath the calyx lobe and can be observed at harvest upon lifting the calyx. ABS infects persimmon fruit through these small wounds under the sepals (calyx) of the fruit and/or directly into the fruit cuticle (Palou *et al.*, 2009, 2012, 2015; Kobiler *et al.*, 2011) during rainy periods. In general, *Alternaria* infections remain quiescent until harvest time, the disease then develops slightly during storage at 0°C and spreads further during shelf life when black spots become apparent during postharvest handling (Prusky *et al.*, 1981). This disease must be controlled by applying antifungal treatments or resistance inducers in both the field and during the postharvest period. Preharvest chemical treatments including fungicides and GA_3 sprays (Pérez *et al.*, 1996) can help to reduce this problem. The most effective current postharvest treatments are modified atmospheres (~30% CO_2) and sodium triclosan or hydrochloric acid baths, either alone or combined with the fungicide prochloraz (Prusky *et al.*, 2001, 2006; Kobiler *et al.*, 2011). The recommended chlorine treatment in 'Triumph', a highly susceptible cultivar, is effective at controlling black spot for up to 2 months; however, the decay incidence increases significantly the longer the fruit is stored (Prusky *et al.*, 2001).

Botrytis rot caused by *Botrytis cinerea*, a wound pathogen, is also a significant problem, especially in cool years. It develops on damaged tissue during storage reaching a full expression during retail handling. *Botrytis* rot can develop large lesions on fruit that was punctured by the apex spike or

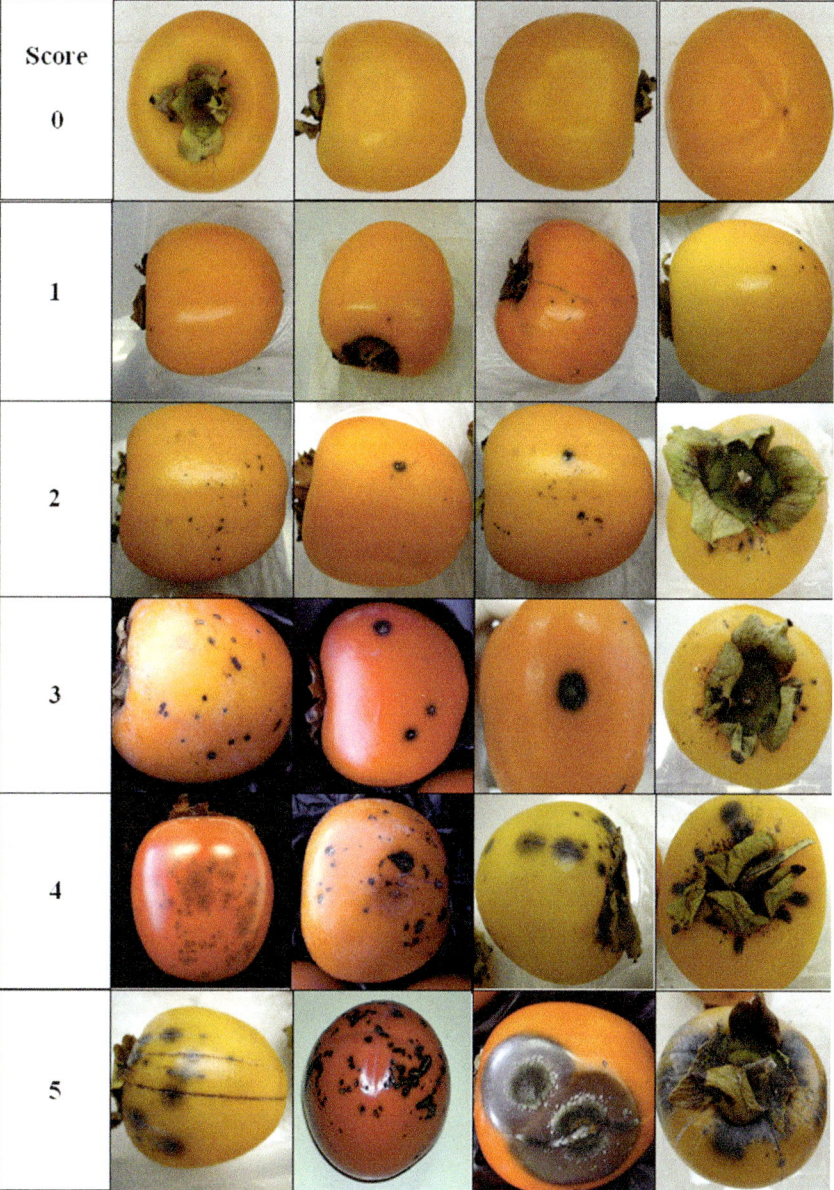

Fig. 4.25. *Alternaria* black spot (ABS) severity index. Photo courtesy of Dr. Luis Palou, Instituto Valenciano de Investigaciones Agrarias, Valencia, Spain.

stylar end from other fruit in the picking bag or bin. The stylar end can be infected during bloom and the rot can develop in these puncture wounds during storage. The symptoms noted during cold storage consist of very soft lesions of varying sizes, which discolor fruit skin. Lesions appear from below the calyx and spread over the peduncle region and are also found occasionally on other fruit areas. Infected fruits that touch adjacent fruits can cause rotting nests. Hot water postharvest treatments have proven to be effective for controlling the disease during long storage periods (Woolf *et al.*, 2008).

Persimmon fruits are also susceptable to peduncular mold caused by several fungi. One of them is known as *Pestalotiopsis clavispora*, which causes a dry mold that commences below the calyx and spreads all over the peduncular region. In some cases, symptoms can be seen in other fruit areas. In central areas of lesions, a white-colored cotton-like fungal mycelium develops (Palou *et al.*, 2009). Another peduncular mold type that has been recently detected in persimmon fruits in the Mediterranean region is caused by *Lasiodiplodia theobromae* (Palou *et al.*, 2013a). This pathogen produces a cotton-like white mycelium on soft, irregular light-brown-colored lesions, which gradually darken when they spread from the calyx to the rest of the fruit. Symptoms are also observed, but less frequently, in the equatorial and stylar fruit regions. These diseases commence mainly in the field from latent infections, and their incidence does not tend to be high and no specific control measures are required.

Anthracnose, which is caused by *Colletotrichum gloeosporioides* and by *Colletotrichum acutatum*, may also affect persimmon fruits. Lesions are rounded and their color varies from dark brown to black (Palou *et al.*, 2013b). They appear more frequently in equatorial fruit regions but can also be displayed in the peduncular region. *Colletotrichum horii* also causes anthracnose symptoms in the field and affects young shoots and fruits even before they are ripe for commercial use. Anthracnose caused by *C. horii* is one of the main diseases in leading persimmon production countries like Japan, China, South Korea, and New Zealand (Weir and Johnston, 2010; Kwon *et al.*, 2013).

Other fungi that can affect persimmon, but to a lesser extent, are *Penicillium expansum, Rhizopus stolonifer, Cladosporium* spp., *Trichoderma* spp., and *Mucor piriformis* (Crisosto, 2004; Kwon and Park, 2004; Palou *et al.*, 2009).

Utilization

Non-astringent persimmons are eaten fresh as a dessert or may be consumed dried or candied and are seldom used in cooking. Astringent persimmons such as 'Triumph' and 'Rojo Brillante' can be used as non-astringent and consumed fresh after treatment with 80% CO_2 for 24 h at

20°C. Dried persimmons require an intensive preparation such as peeling and drying. Additional persimmon processing includes dried powders from purees to prepare traditional sherbets, and jams. Persimmon puree is used in baked goods, puddings, and custards.

Retail Outlet Display Considerations

Cold-table display is recommended.

Suitability as a Fresh-cut Product

Non-astringent persimmon cultivars can be prepared as fresh-cut wedges or slices (M.B. Perez-Gago, Spain, 2019, personal communication). It has been reported that the shelf life of 'Fuyu' persimmon slices was 7 days in air and 8 days in CA of 2% O_2 and 12% CO_2 at 5°C. A longer shelf life can be expected at 0–2°C. Protecting slices from ethylene helps firmness retention.

Special Research Needs

Research needs include:

- the search for a large size, flavorful PCNA cultivar;
- validation of ethanol and carbon dioxide combined treatment to improve astringency removal at low temperatures;
- use of breeding programs to add flavor diversity, and remove CI and ethylene sensitivity;
- generation of data on the health benefits of consuming persimmons using human studies;
- more studies to find the optimal CA conditions for different persimmon cultivars;
- studies to optimize postharvest technology (MAP, 1-MCP, others) to extend postharvest life; and
- studies to improve packing lines to reduce impact damage and protect persimmon quality.

References

Agustí, M., Juan, M., Yagüe, B., Mesejo, C., Martínez-Fuentes, A. *et al.* (2003) Tratamientos para retrasar la maduración del fruto del caqui (*Diospyros kaki* L.). *Comunidad Valenciana Agraria* 24, 27–33.

Argenta, L.C., Vieira, M.J. Scolaro, A.M.T. and Tomazini, A.M. (2009) Conservação da qualidade de caqui 'Fuyu' em ambiente refrigerado pela combinação de

1-MCP e atmosfera modificada. *Revista Brasileira de Fruticultura* 31(2), 323–333. DOI: 10.1590/S0100-29452009000200006.

Arnal, L. and Del Río, M.A. (2004) Effect of cold storage and removal astringency on quality of persimmon fruit (*Diospyros kaki*, L.) cv. Rojo Brillante. *Food Science and Technology International* 10(3), 179–185. DOI: 10.1177/1082013204044824.

Arnal, L., Besada, C., Navarro, P. and Salvador, A. (2008) Effect of controlled atmospheres on maintaining quality of persimmon fruit cv. 'Rojo Brillante'. *Journal of Food Science* 73(1), S26–S30. DOI: 10.1111/j.1750-3841.2007.00602.x.

Ben-Arie, R., Zutkhi, Y., Sonego, L. and Klein, J. (1991) Modified atmosphere packaging for long-term storage of astringent persimmons. *Postharvest Biology and Technology* 1(2), 169–179. DOI: 10.1016/0925-5214(91)90009-Z.

Ben-Arie, R., Zhou, H.W., Sonego, L. and Zutkhi, Y. (1997) Plant growth regulator effects on the storage and shelf-life 'Triumph' persimmons. *Acta Horticulturae* 436, 243–250.

Besada, C., Arnal, L. and Salvador, A. (2008a) Improving storability of persimmon cv. Rojo Brillante by combined use of preharvest and postharvest treatments. *Postharvest Biology and Technology* 50(2-3), 169–175. DOI: 10.1016/j.postharvbio.2008.05.013.

Besada, C., Salvador, A., Arnal, L. and Martínez-Jávega, J.M. (2008b) Hot water treatment for chilling injury reduction of astringent 'Rojo Brillante' persimmon at different maturity stages. *HortScience* 43(7), 2120–2123. DOI: 10.21273/HORTSCI.43.7.2120.

Besada, C., Jackman, R.C., Olsson, S. and Woolf, A.B. (2010a) Response of 'Fuyu' persimmons to ethylene exposure before and during storage. *Postharvest Biology and Technology* 57(2), 124–131. DOI: 10.1016/j.postharvbio.2010.03.002.

Besada, C., Salvador, A., Arnal, L. and Martínez-Jávega, J.M. (2010b) Optimization of the duration of deastringency treatment depending on persimmon maturity. *Acta Horticulturae* 858,69–74. DOI: 10.17660/ActaHortic.2010.858.7.

Besada, C., Novillo, P., Navarro, P. and Salvador, A. (2014) Effect of a low oxygen atmosphere combined with 1-MCP pretreatment on preserving the quality of 'Rojo Brillante' and 'Triumph' persimmon during cold storage. *Scientia Horticulturae* 179, 51–58. DOI: 10.1016/j.scienta.2014.09.015.

Besada, C., Llorca, E., Novillo, P., Hernando, I. and Salvador, A. (2015) Short-term high CO_2 treatment alleviates chilling injury of persimmon cv. Fuyu by preserving the parenchyma structure. *Food Control* 51, 163–170. DOI: 10.1016/j.foodcont.2014.11.013.

Besada, C., Gil, R., Bonet, L., Quiñones, A., Intrigliolo, D. *et al.* (2016) Chloride stress triggers maturation and negatively affects the postharvest quality of persimmon fruit. Involvement of calyx ethylene production. *Plant Physiology and Biochemistry* 100, 105–112. DOI: 10.1016/j.plaphy.2016.01.006.

Besada, C., Novillo, P., Navarro, P. and Salvador, A. (2018) Causes of flesh browning in persimmon – a review. *Acta Horticulturae* 1195(1195), 203–210. DOI: 10.17660/ActaHortic.2018.1195.32.

Brackmann, A., Mazaro, S.M. and Saquet, A.A. (1997) Cold storage of persimmons (*Diospyros kaki* L.) cultivars Fuyu and Rama Forte. *Ciencia-Rural* 27, 561–565.

Brackmann, A., Vilela, J.A., Silveira, A.C., Steffens, C. and Sestari, I. (2006) Storage conditions of 'Fuyu' persimmon. *Revista Brasileira de Agrociência* 12, 183–186.

Burmeister, D.M., Ball, S., Green, S. and Woolf, A.B. (1997) Interaction of hot water treatments and controlled atmosphere storage on quality of 'Fuyu' persimmons. *Postharvest Biology and Technology* 12(1), 71–81. DOI: 10.1016/S0925-5214(97)00029-X.

Candir, E.E., Ozdemir, A.E., Kaplankiran, M. and Toplu, C. (2009) Physico-chemical changes during growth of persimmon fruits in the East Mediterranean climate region. *Scientia Horticulturae* 121(1), 42–48. DOI: 10.1016/j.scienta.2009.01.009.

Collins, R.J. and Tisdell, J.S. (1995) The influence of storage time and temperature on chilling injury in Fuyu and Suruga persimmon (*Diospyros* kaki L.) grown in subtropical Australia. *Postharvest Biology and Technology* 6(1-2), 149–157. DOI: 10.1016/0925-5214(94)00046-U.

Crisosto, C. (2004) Persimmon. In: Gross, K.C., Wang, C.Y. and Salveit, M. (eds) *The Commercial Storage of Fruits, Vegetables, and Florist and Nursery Stocks. Agriculture Handbook 66.* United States Department of Agriculture Agricultural Research Service, Washington, DC, pp. 487–489. Available at: https://www.ars.usda.gov/ARSUserFiles/oc/np/CommercialStorage/CommercialStorage.pdf (accessed 13 July 2020).

Crisosto, C.H., Mitcham, E.J. and Kader, A.A. (1995) Produce facts: persimmons. *Perishables Handling Quarterly* 84, 19–20.

Daniell, R., Girardi, C.L., Parussolo, A., Ferri, V.C. and Rombaldi, C.V. (2002) Effect of the application of gibberellic acid and calcium chloride in the retardation of harvest and conservability of persimmon, Fuyu. *Revista Brasileira de Fruticultura* 24, 44–48.

Del Bubba, M., Giordani, E., Pippucci, L., Cincinelli, A., Checchini, L. *et al.* (2009) Changes in tannins, ascorbic acid and sugar content in astringent persimmons during on-tree growth and ripening and in response to different postharvest treatments. *Journal of Food Composition and Analysis* 22(7-8), 668–677. DOI: 10.1016/j.jfca.2009.02.015.

Ebert, G. and Gross, J. (1985) Carotenoid changes in the peel of ripening persimmon (*Diospyros kaki*) cv Triumph. *Phytochemistry* 24(1), 29–32. DOI: 10.1016/S0031-9422(00)80801-8.

FAOSTAT (2019) Digital Report - the State of Food and Agriculture. Food and Agricultural Organization of the United Nations, Rome. Available at: http://www.fao.org/statistics/en/ (accessed 11 May 2020).

Gazit, S. and Adato, I. (1972) Effect of carbon dioxide atmosphere on the course of astringency disappearance of persimmon (*Diospyros kaki* Linn.) fruits. *Journal of Food Science* 37(6), 815–817. DOI: 10.1111/j.1365-2621.1972.tb03676.x.

Giordani, E., Doumett, S., Nin, S. and Del Bubba, M. (2011) Selected primary and secondary metabolites in fresh persimmon (*Diospyros kaki* Thunb.): a review of analytical methods and current knowledge of fruit composition and health benefits. *Food Research International* 44(7), 1752–1767. DOI: 10.1016/j.foodres.2011.01.036.

Glucina, P.G. (1987) Calyx separation: a physiological disorder of persimmons. *Orchardist of New Zealand* 60, 161–163.

Gorinstein, S., Bartnikowska, E., Kulasek, G., Zemser, M. and Trakhtenberg, S. (1998) Dietary persimmon improves lipid metabolism in rats fed diets containing cholesterol. *The Journal of Nutrition* 128(11), 2023–2027. DOI: 10.1093/jn/128.11.2023.

Gorinstein, S., Leontowicz, H., Leontowicz, M., Jesion, I., Namiesnik, J. *et al.* (2011) Influence of two cultivars of persimmon on atherosclerosis indices in rats fed cholesterol-containing diets: investigation *in vitro* and *in vivo*. *Nutrition* 27(7-8), 838–846. DOI: 10.1016/j.nut.2010.08.015.

Grant, T.M., Macrae, E.A. and Redgwell, R.J. (1992) Effect of chilling injury on physicochemical properties of persimmon cell walls. *Phytochemistry* 31(11), 3739–3744. DOI: 10.1016/S0031-9422(00)97519-8.

Ikegami, A., Kitajima, A. and Yonemori, K. (2005) Inhibition of flavonoid biosynthetic gene expression coincides with loss of astringency in pollination-constant, non-astringent (PCNA)-type persimmon fruit. *The Journal of Horticultural Science and Biotechnology* 80(2), 225–228. DOI: 10.1080/14620316.2005.11511921.

Itamural, H., Kitamura, T., Taira, S., Harada, H., Ito, N. *et al.* (1991) Relationship between fruit softening, ethylene production and respiration in Japanese persimmon 'Hiratanenashi'. *Engei Gakkai zasshi* 60(3), 695–701. DOI: 10.2503/jjshs.60.695.

Ito, S. (1971) The persimmon. In: Hulme, A.C. (ed.) *The Biochemistry of Fruits and Their Products.* Academic Press, New York, pp. 281–301.

Ito, S. (1980) Persimmon. In: Nagy, S. and Shaw, P.E. (eds) *Tropical and Subtropical Fruits: Composition, Properties and Uses.* AVI, Westport, Connecticut, pp. 442–468.

Khademi, O., Besada, C., Mostofi, Y. and Salvador, A. (2014) Changes in pectin methylesterase, polygalacturonase, catalase and peroxidase activities associated with alleviation of chilling injury in persimmon by hot water and 1-MCP treatments. *Scientia Horticulturae* 179, 191–197. DOI: 10.1016/j.scienta.2014.09.028.

Kim, Y.K. and Lee, J.M. (2005) Extension of storage and shelf-life of sweet persimmon with 1-MCP. *Acta Horticulturae* 685,165–175. DOI: 10.17660/ActaHortic.2005.685.19.

Kim, Y.H., Lim, S.C., Youn, C.K., Yoon, T. and Kim, T.S. (2004) Effect of ethephon on fruit quality and maturity of 'Tone Wase' astringent persimmons (*Diospyros kaki* L.). *Acta Horticulturae* 653,187–191. DOI: 10.17660/ActaHortic.2004.653.26.

Kobiler, I., Akerman, M., Huberman, L. and Prusky, D. (2011) Integration of pre- and postharvest treatments for the control of black spot caused by *Alternaria alternata* in stored persimmon fruit. *Postharvest Biology and Technology* 59(2), 166–171. DOI: 10.1016/j.postharvbio.2010.08.009.

Krammes, J.G., Argenta, L.C. and Vieira, M.J. (2006) Influences of 1-methylcyclopropene on quality of persimmon fruit cv. 'Fuyu' after cold storage. *Acta Horticulturae* 727,513–518. DOI: 10.17660/ActaHortic.2006.727.63.

Kubo, Y., Nakano, R. and Inaba, A. (2003) Cloning of genes encoding cell wall modifying enzymes and their expression in persimmon fruit. *Acta Horticulturae* 601,49–55. DOI: 10.17660/ActaHortic.2003.601.5.

Kwon, J.-H. and Park, C.-S. (2004) Ecology of disease outbreak of circular leaf spot of persimmon and inoculum dynamics of *Mycosphaerella nawae*. *Research in Plant Disease* 10(4), 209–216. DOI: 10.5423/RPD.2004.10.4.209.

Kwon, J.-H., Kim, J., Choi, O., Gang, G.-H., Han, S. *et al.* (2013) Anthracnose caused by *Colletotrichum horii* on sweet persimmon in Korea: dissemination of conidia and disease development. *Journal of Phytopathology* 161(7-8), 497–502. DOI: 10.1111/jph.12096.

Lay-Yee, M., Ball, S., Forbes, S.K. and Woolf, A.B. (1997) Hot-water treatment for insect disinfestation and reduction of chilling injury of 'Fuyu' persimmon. *Postharvest Biology and Technology* 10(1), 81–87. DOI: 10.1016/S0925-5214(97)87277-8.

Lee, Y.M., Jang, S.J. and Lee, Y.J. (1997) Effect of preharvest application of MGC-140 and GA3 on the storability of 'Fuyu' persimmon (*Diospyros kaki* L). *Journal of the Korean Society for Horticultural Science* 38, 157–161.

Luo, Z.S. and Xi, Y.F. (2005) Effect of storage temperature on physiology and ultra-structure of persimmon fruit. *Journal of Zhejiang University Agriculture and Life Sciences* 31, 195–198.

Macarthur, E. (2003) Persimmon Growing in New South Wales. Agfact H3.117, 3rd edition. NSW Agriculture, NSW Government, Australia. Available at: https://www.dpi.nsw.gov.au/__data/assets/pdf_file/0006/119517/persimmon-growing.pdf (accessed 11 May 2020).

MacRae, E.A. (1987) Development of chilling injury in New Zealand grown 'Fuyu' persimmon during storage. *New Zealand Journal of Experimental Agriculture* 15(3), 333–344. DOI: 10.1080/03015521.1987.10425579.

MacRae, E.A. and New Zealand Division of Horticulture and Processing (1987) *Storage and shelf life of Fuyu and flat Fuyu persimmon in New Zealand, 1984–1986.* Division of Horticulture and Processing, Dept. of Scientific and Industrial Research, Auckland, New Zealand.

Matsuo, T. and Ito, S. (1977) On mechanisms of removing astringency in persimmon fruits by carbon dioxide treatment I. Some properties of the two processes in the de-astringency. *Plant and Cell Physiology* 18, 17–25.

Matsuo, T. and Ito, S. (1982) A model experiment for de-astringency of persimmon fruit with high carbon dioxide treatment: *in vitro* gelation of kaki-tannin by reacting with acetaldehyde. *Agricultural and Biological Chemistry* 46(3), 683–689. DOI: 10.1271/bbb1961.46.683.

Mowat, A.D. and George, A.P. (1996) Environmental physiology of persimmons. In: Schaffer, B. and Andersen, P. (eds) *Handbook of Environmental Physiology of Fruit Crops.* CRC Press, Boca Raton, Florida, pp. 195–202.

Munera, S., Besada, C., Blasco, J., Cubero, S., Salvador, A. *et al.* (2017) Astringency assessment of persimmon by hyperspectral imaging. *Postharvest Biology and Technology* 125, 35–41. DOI: 10.1016/j.postharvbio.2016.11.006.

Nakano, R., Yonemori, K., Sugiura, A. and Kataoka, I. (1997) Effect of gibberellic acid and abscisic acid on fruit respiration in relation to final swell and maturation in persimmon. *Acta Horticulturae* 436,203–214. DOI: 10.17660/ActaHortic.1997.436.23.

Niikawa, T., Suzuki, T., Ozeki, T., Kato, M. and Ikoma, Y. (2007) Characteristics of carotenoid accumulation during maturation of the Japanese persimmon 'Fuyu'. *Horticultural Research* 6(2), 251–256. DOI: 10.2503/hrj.6.251.

Novillo, P., Besada, C., Gil, R. and Salvador, A. (2013) Fruit quality and response to deastringency treatment of eight persimmon varieties cultivated under Spanish growing conditions. *Acta Horticulturae* 996,437–442. DOI: 10.17660/ActaHortic.2013.996.63.

Novillo, P., Salvador, A., Llorca, E., Hernando, I. and Besada, C. (2014) Effect of CO_2 deastringency treatment on flesh disorders induced by mechanical damage in persimmon. Biochemical and microstructural studies. *Food Chemistry* 145, 454–463. DOI: 10.1016/j.foodchem.2013.08.054.

Novillo, P., Besada, C., Tian, L., Bermejo, A. and Salvador, A. (2015a) Nutritional composition of ten persimmon cultivars in the 'ready-to-eat crisp' stage. Effect of deastringency treatment. *Food and Nutrition Sciences* 06(14), 1296–1306. DOI: 10.4236/fns.2015.614135.

Novillo, P., Salvador, A., Navarro, P. and Besada, C. (2015b) Involvement of the redox system in chilling injury and its alleviation by 1-methylcyclopropene in 'Rojo Brillante' persimmon. *HortScience* 50(4), 570–576. DOI: 10.21273/HORTSCI.50.4.570.

Novillo, P., Salvador, A., Navarro, P. and Besada, C. (2015c) Sensitivity of astringent and non-astringent persimmon cultivars to flesh disorders induced by mechanical damage. *Acta Horticulturae* 1079,605–610. DOI: 10.17660/ActaHortic.2015.1079.82.

Orihuel-Iranzo, B., Miranda, M., Zacarías, L. and Lafuente, M.T. (2010) Temperature and ultra low oxygen effects and involvement of ethylene in chilling injury of 'Rojo Brillante' persimmon fruit. *Food Science and Technology International* 16(2), 159–167. DOI: 10.1177/1082013209353221.

Palou, L., Montesinos-Herrero, C., Guardado, A., Besada, C. and Del Río, M.A. (2009) Fungi associated with postharvest decay of persimmon in Spain. *Acta Horticulturae* 833,275–280. DOI: 10.17660/ActaHortic.2009.833.44.

Palou, L., Taberner, V., Guardado, A. and Montesinos-Herrero, C. (2012) First report of *Alternaria alternata* causing postharvest black spot of persimmon in Spain. *Australasian Plant Disease Notes* 7(1), 41–42. DOI: 10.1007/s13314-012-0043-0.

Palou, L., Montesinos-Herrero, C., Besada, C. and Taberner, V. (2013a) Postharvest fruit rot of persimmon (*Diospyros kaki*) in Spain caused by *Lasiodiplodia theobromae* and *Neofusicoccum* spp. *Journal of Phytopathology* 161(9), 625–631. DOI: 10.1111/jph.12111.

Palou, L., Montesinos-Herrero, C., Tarazona, I. and Taberner, V. (2013b) Postharvest anthracnose of persimmon fruit caused by *Colletotrichum gloeosporioides* first reported in Spain. *Plant Disease* 97(5), 691–691. DOI: 10.1094/PDIS-11-12-1044-PDN.

Palou, L., Montesinos-Herrero, C., Tarazona, I., Besada, C. and Taberner, V. (2015) Incidence and etiology of postharvest fungal diseases of persimmon (*Diospyros kaki* Thunb. cv. Rojo Brillante) in Spain. *Plant Disease* 99(10), 1416–1425. DOI: 10.1094/PDIS-01-15-0112-RE.

Park, Y.-M. and Lee, Y.-J. (2008) Induction of modified atmosphere-related browning disorders in 'Fuyu' persimmon fruit. *Postharvest Biology and Technology* 47(3), 346–352. DOI: 10.1016/j.postharvbio.2007.08.006.

Pérez-Munuera, I., Quiles, A., Larrea, V., Arnal, L., Besada, C. *et al.* (2009a) Microstructure of persimmon treated by hot water to alleviate chilling injury. *Acta Horticulturae* 883,251–256. DOI: 10.17660/ActaHortic.2009.833.40.

Pérez, A., Ben-Arie, R., Dinoor, A., Genizi, A. and Prusky, D. (1996) Prevention of black spot disease in persimmon fruit by gibberellic acid and iprodione treatments. *Phytopathology* 85(2), 221–225. DOI: 10.1094/Phyto-85-221.

Pérez-Munuera, I., Hernando, I., Larrea, V., Besada, C., Arnal, L. *et al.* (2009b) Microstructural study of chilling injury alleviation by 1-methylcyclopropene in persimmon. *HortScience* 44(3), 742–745. DOI: 10.21273/HORTSCI.44.3.742.

Pesis, E., Levi, A. and Ben-Arie, R. (1988) Role of acetaldehyde production in the removal of astringency from persimmon fruits under various modified atmospheres. *Journal of Food Science* 53(1), 153–156. DOI: 10.1111/j.1365-2621.1988.tb10197.x.

Prusky, D., Ben-Arie, R. and Guelfat-Reich, S. (1981) Etiology and histology of Alternaria rot of persimmon fruits. *Phytopathology* 71(11), 1124–1128. DOI: 10.1094/Phyto-71-1124.

Prusky, D., Eshel, D., Kobiler, I., Yakoby, N., Beno-Moualem, D. *et al.* (2001) Postharvest chlorine treatments for the control of the persimmon black spot disease caused by *Alternaria alternata*. *Postharvest Biology and Technology* 22(3), 271–277. DOI: 10.1016/S0925-5214(01)00084-9.

Prusky, D., Kobiler, I., Akerman, M. and Miyara, I. (2006) Effect of acidic solutions and acidic prochloraz on the control of postharvest decay caused by *Alternaria alternata* in mango and persimmon fruit. *Postharvest Biology and Technology* 42(2), 134–141. DOI: 10.1016/j.postharvbio.2006.06.001.

Salvador, A., Arnal, L., Monterde, A. and Cuquerella, J. (2004) Reduction of chilling injury symptoms in persimmon fruit cv. 'Rojo Brillante' by 1-MCP. *Postharvest Biology and Technology* 33(3), 285–291. DOI: 10.1016/j.postharvbio.2004.03.005.

Salvador, A., Arnal, L., Monterde, A. and Martínez-Jávega, J.M. (2005) Influence of ripening stage at harvest on chilling injury symptoms of persimmon cv. Rojo Brillante stored at different temperatures. *Food Science and Technology International* 11(5), 359–365. DOI: 10.1177/1082013205057941.

Salvador, A., Arnal, L., Carot, J.M., Carvalho, C.P. and Jabaloyes, J.M. (2006) Influence of different factors on firmness and color evolution during the storage of persimmon cv. 'Rojo Brillante'. *Journal of Food Science* 71(2), S169–S175. DOI: 10.1111/j.1365-2621.2006.tb08921.x.

Salvador, A., Arnal, L., Besada, C., Larrea, V., Quiles, A. *et al.* (2007) Physiological and structural changes during ripening and deastringency treatment of persimmon fruit cv. 'Rojo Brillante'. *Postharvest Biology and Technology* 46(2), 181–188. DOI: 10.1016/j.postharvbio.2007.05.003.

Salvador, A., Arnal, L., Besada, C., Larrea, V., Hernando, I. *et al.* (2008) Reduced effectiveness of the treatment for removing astringency in persimmon fruit when stored at 15°C: physiological and microstructural study. *Postharvest Biology and Technology* 49(3), 340–347. DOI: 10.1016/j.postharvbio.2008.01.015.

Santos-Buelga, C. and Scalbert, A. (2000) Proanthocyanidins and tannin-like compounds–nature, occurrence, dietary intake and effects on nutrition and health. *Journal of the Science of Food and Agriculture* 80(7), 1094–1117. DOI: 10.1002/(SICI)1097-0010(20000515)80:7<1094::AID-JSFA569>3.0.CO;2-1.

Senter, S.D., Chapman, G.W., Forbus, W.R. and Payne, J.A. (1991) Sugar and non-volatile acid composition of persimmons during maturation. *Journal of Food Science* 56(4), 989–991. DOI: 10.1111/j.1365-2621.1991.tb14623.x.

Sugiura, A. and Tomana, T. (1983) Relationships of ethanol production by seeds of different types of Japanese persimmons and their tannin content [*Diospyros kaki*, pollination, astringency]. *HortScience* 18, 319–321.

Taira, S., Itamura, H., Abe, K. and Watanabe, S. (1989) Comparison of the characteristics of removal of astringency in two Japanese persimmon cultivars, 'Denkuro' and 'Hiratanenashi'. *Journal of the Japanese Society for Horticultural Science* 58, 319–325.

Taira, S., Itamura, H., Abe, K., Oor, K. and Watanabe, S. (1990) Effect of harvest maturity on removal of astringency in Japanese persimmon (*Diospyros kaki* Thunb.), 'Hratanenashi' fruits. *Journal of the Japanese Society for Horticultural Science* 58, 813–818.

Taira, S., Oba, S. and Watanabe, S. (1992a) Removal of astringency from 'Hiratanenashi' persimmon fruit with a mixture of ethanol and carbon dioxide. *Journal of the Japanese Society for Horticultural Science* 61(2), 437–443. DOI: 10.2503/jjshs.61.437.

Taira, S., Satoh, I. and Watanabe, S. (1992b) Relationship between differences in the ease of removal of astringency among fruits of Japanese persimmon (*Diospyros kaki* Thunb.) and their ability to accumulate ethanol and acetaldehyde. *Journal of the Japanese Society for Horticultural Science* 60(4), 1003–1009. DOI: 10.2503/jjshs.60.1003.

Taira, S., Ono, M. and Matsumoto, N. (1997) Reduction of persimmon astringency by complex formation between pectin and tannins. *Postharvest Biology and Technology* 12(3), 265–271. DOI: 10.1016/S0925-5214(97)00064-1.

Taira, S., Matsumoto, N. and Ono, M. (1998) Accumulation of soluble and insoluble tannins during fruit development in nonastringent and astringent persimmon. *Journal of the Japanese Society for Horticultural Science* 67(4), 572–576. DOI: 10.2503/jjshs.67.572.

Tanaka, T., Takahashi, R., Kouno, I. and Nonaka, G.I. (1994) Chemical evidence for the de-astringency (insolubilization of tannins) of persimmon fruit. *Journal of the Chemical Society, Perkin Transactions* 1(20), 3013–3022. DOI: 10.1039/p19940003013.

Tessmer, M.A., Besada, C., Hernando, I., Appezzato-da-Glória, B., Quiles, A. *et al.* (2016) Microstructural changes while persimmon fruits mature and ripen. Comparison between astringent and non-astringent cultivars. *Postharvest Biology and Technology* 120, 52–60. DOI: 10.1016/j.postharvbio.2016.05.014.

Tsviling, A., Nerya, O., Gizis, A., Sharabi-Nov, A. and Ben-Arie, R. (2003) Extending the shelf-life of 'Triumph' persimmons after storage, with 1-MCP. *Acta Horticulturae* 599,53–58. DOI: 10.17660/ActaHortic.2003.599.4.

Weir, B.S. and Johnston, P.R. (2010) Characterisation and neotypification of *Gloeosporium kaki* Hori as *Colletotrichum horii* nom. nov. *Mycotaxon* 111, 209–219.

Wills, R., McGlasson, B., Graham, D. and Joyce, D. (1998) *Postharvest: an Introduction to the Physiology & Handling of Fruit, Vegetables & Ornamentals*, 4th edn. CAB International, Wallingford, UK.

Woolf, A.B. and Ben-Arie, R. (2011) Persimmon (*Diospyros kaki* L.). In: Yahia, E.M. (ed.) *Postharvest Biology and Technology of Tropical and Subtropical Fruits*. Woodhead Publishing, Cambridge, UK, pp. 166–193.

Woolf, A.B., MacRae, E.A., Spooner, K.J. and Redgwell, R.J. (1997a) Changes to physical properties of the cell wall and polyuronides in response to heat treatment of 'Fuyu' persimmon that alleviate chilling injury. *Journal of the American Society for Horticultural Science* 122(5), 698–702. DOI: 10.21273/JASHS.122.5.698.

Woolf, A.B., Ball, S., Spooner, K.J., Lay-Yee, M., Ferguson, I.B. *et al.* (1997b) Reduction of chilling injury in the sweet persimmon 'Fuyu' during storage by dry air heat treatments. *Postharvest Biology and Technology* 11(3), 155–164. DOI: 10.1016/S0925-5214(97)00024-0.

Woolf, A., Jackman, R., Olsson, S., Manning, M., Rheinlander, P. *et al.* (2008) Meeting consumer requirements from a New Zealand perspective. *Advances in Horticultural Science* 22, 274–280.

Wright, K.P. and Kader, A.A. (1997) Effect of slicing and controlled-atmosphere storage on the ascorbate content and quality of strawberries and persimmons. *Postharvest Biology and Technology* 10(1), 39–48. DOI: 10.1016/S0925-5214(96)00061-0.

Yamada, M., Taira, S., Ohtsuki, M., Sato, A., Iwanami, H. *et al.* (2002) Varietal differences in the ease of astringency removal by carbon dioxide gas and ethanol vapor treatments among Oriental astringent persimmons of Japanese and Chinese origin. *Scientia Horticulturae* 94(1-2), 63–72. DOI: 10.1016/S0304-4238(01)00367-3.

Yokozawa, T., Kim, Y.A., Kim, H.Y., Lee, Y.A. and Nonaka, G.-I. (2007) Protective effect of persimmon peel polyphenol against high glucose-induced oxidative stress in LLC-PK(1) cells. *Food and Chemical Toxicology* 45(10), 1979–1987. DOI: 10.1016/j.fct.2007.04.018.

Yonemori, K. and Matsushima, J. (1985) Property of development of the tannin cells in non-astringent type fruits of Japanese persimmon (*Diospyros kaki*) and its relationship to natural deastringency. *Journal of the Japanese Society for Horticultural Science* 54(2), 201–208. DOI: 10.2503/jjshs.54.201.

Yonemori, K. and Suzuki, Y. (2009) Differences in three-dimensional distribution of tannin cells in flesh tissue between astringent and non-astringent type persimmon. *Acta Horticulturae* 833, 119–124. DOI: 10.17660/ActaHortic.2009.833.18.

Yonemori, K., Ikegami, A., Kanzaki, S. and Sugiura, A. (2003) Unique features of tannin cells in fruit of pollination constant non-astringent persimmons. *Acta Horticulturae* 601, 31–35. DOI: 10.17660/ActaHortic.2003.601.3.

Zhang, Y., Rao, J., Sun, Y. and Li, S. (2010) Reduction of chilling injury in sweet persimmon fruit by 1-MCP. *Acta Horticulturae Sinica* 37, 547–552.

Zheng, G.H. and Sugiura, A. (1990) Changes in sugar composition in relation to invertase activity in the growth and ripening of persimmon (*Diospyros kaki*) fruits. *Journal of the Japanese Society for Horticultural Science* 59(2), 281–287. DOI: 10.2503/jjshs.59.281.

Zhou, C., Zhao, D., Sheng, Y., Tao, J. and Yang, Y. (2011) Carotenoids in fruits of different persimmon cultivars. *Molecules* 16(1), 624–636. DOI: 10.3390/molecules16010624.

Pistachio

5

Carlos H. Crisosto* and Louise Ferguson
University of California, Davis, California, USA

Scientific Name, Origin and Current Areas of Production

All pistachio trees belong to the genus *Pistacia* which comprises both male and female commercial varieties, rootstocks, and wild relatives. The genus *Pistacia* belongs to the family *Anacardiaceae*, which also includes poison ivy, poison oak, cashew, sumac, pepper tree, and mango; all of them exude latex or resin. The word pistachio was taken from the Zendor Avestan (ancient Persian language) word pista-pistak and is a cognate to the modern Persian word peste. According to Dioskurides the pistachio is derived from pissa (means resin) and aklomai (means to heal) (i.e. it is a plant with healthy resin) (Kashaninejad and Tabil, 2011). The common name of pistachio in different languages is: peste (Persian), pistache (French), pistazie (German), pistacchio (Italian), pistacho (Spanish) pista (Indian), pistasch (Swedish), fustuq (Arabic) and pisutachio (Japanese) (Kashaninejad and Tabil, 2011).

Pistacia species are native to latitudes between 30° and 40° including Asia Minor (now Turkey), Iran, Syria, Lebanon, the Caucasus, southern Russia, and Afghanistan. It probably developed in interior desert areas because it requires long, hot summers for fruit maturation, is drought and salt tolerant, and has a high winter chilling requirement. Recently, cultivars with low chilling requirements, early harvest, low blank and high split and size such as 'Kerman' are being produced. These new cultivars assure high quality production in areas with low chilling and in years with low chilling accumulation (Parfitt *et al.*, 2016). Also, new males have been released to provide satisfactory pollination under those conditions (Parfitt *et al.*, 2016). Archeologists have found evidence at a dig site at Jarmo, in north-eastern Iraq, indicating that pistachio nuts were a common food as early as 6750 BC. Nebuchadnezzar, the ancient king of Babylon, had pistachio trees planted in his fabled hanging gardens around the 8th century BC (Kashaninejad and Tabil, 2011). There are records of pistachios growing in Syria in the 1st century BC and some of these nuts were shipped from Syria

*Corresponding author: chcrisosto@ucdavis.edu

to Italy in the 1st century AD and spread throughout the Mediterranean countries. Pistachios were also reported to spread eastward from their center of origin, reaching China around the 10th century AD. *Pistacia vera* L. is the only edible species of the 11 species in the genus. From these areas *P. vera* L. was introduced to Europe at the beginning of the Christian Era, and the United States Department of Agriculture (USDA) Plant Exploration Service introduced the pistachio to the USA in 1890. Fourteen years later, pistachios reached the Plant Introduction Station in Chico, California. However, it took nearly 60 years after the Californian introduction before the first commercial plantings were established in California (Ferguson and Kallsen, 2016).

Currently, pistachios are grown in the Central, San Joaquin and northern Sacramento Valleys with the most productive orchards concentrated in the lower west side of California's San Joaquin Valley where they are exposed to: (i) sufficient winter chill; (ii) sustained summer heat without rain and low relative humidity (RH); (iii) deep, loamy soils with high boron concentrations; and (iv) affordable, available water. Pistachios reached Australia after that. The species used to produce edible nuts, *P. vera*, along with multiple other species of *Pistacia* are still growing wild throughout the desert countries of Asia, Africa, southern Europe, and southern North America. According to the Food and Agriculture Organization of the United Nations (FAO), pistachio nuts are produced commercially in 18 countries on 608,729 ha. The top six pistachio producers are Iran (44.4% of the world's production), followed by the USA (20.97% share), Turkey (14.18% share), Syria (10.05% share), China (7.34%, share) and Greece (1.7% share) (FAOSTAT, 2018). Recently, plantings have increased dramatically in the major traditional producer countries and begun in Australia, Mexico, Argentina, and South Africa.

Composition, Quality and Health Benefits

Pistachio nuts are highly valued for their nutritional, sensory and healthy attributes (Kader *et al.*, 1982; Ghasemi-Varnamkhasti, 2015). In addition to being high in unsaturated fatty acids and low in saturated fatty acids, they are good sources of proteins, dietary fibers, vitamins, minerals, and antioxidant phytochemicals (Rodríguez-Bencomo *et al.*, 2015; Pantano *et al.*, 2016). Nutritionally, pistachio nuts are better than other nuts including peanuts, because they have lower calories and fat content; and are higher in protein, carbohydrates and potassium (Carughi *et al.*, 2019). Pistachio kernel (Fig. 5.1) composition is dependent on cultivar, maturity, and region. Mono- and polyunsaturated fatty acids constitute more than 80% of pistachio fats (Labavitch *et al.*, 1982; Polari *et al.*, 2019). The effect of environmental conditions on biosynthesis of fatty acids and terpenes in pistachio kernels, the main compounds responsible for pistachio nutritional

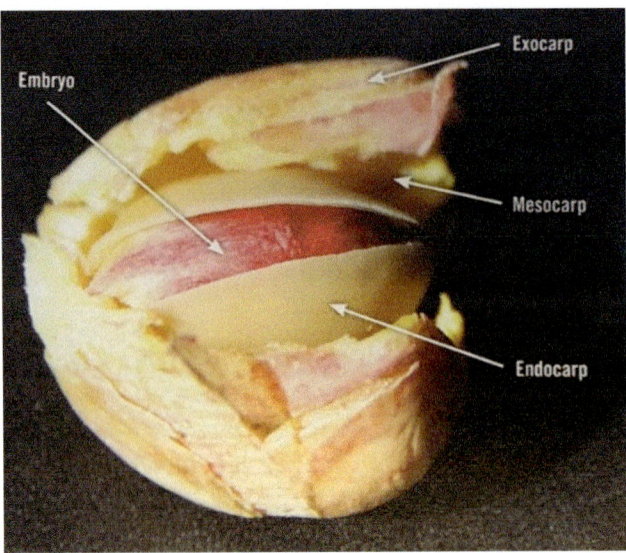

Fig. 5.1. Pistachio nut components. Photo courtesy of Dr. Carlos H. Crisosto.

and sensory characteristics, was demonstrated on cultivars 'Kerman' and 'Golden Hills' growing in California's Central Valley (Polari *et al.*, 2019) and other cultivars (Barla Demirkoz *et al.*, 2018). In general, the main fatty acid in pistachio oil is oleic acid (46–59%), followed by linoleic acid (26–36%), and palmitic acid (11–16%). C16:0, C16:1, C18:2, and C18:3 decreased with harvest time, while C18:1 increased. Alpha-pinene was the most concentrated volatile among the cultivars and locations. It decreased with harvest date for both cultivars at both locations, ranging from 105 mg kg^{-1} to 2,464 mg kg^{-1}. At harvest, 'Golden Hills' and 'Kerman' at Parlier both had higher concentrations of α-pinene than the two cultivars at 'Lost Hills' (Polari *et al.*, 2019). Carbohydrate analysis of pistachio has indicated the predominant sugar to be sucrose followed by raffinose, glucose, fructose, maltose, and stachyose. The globulin fraction is the major protein in the pistachio, contributing about 66% of the total protein. Albumins are second in predominance to globulins, contributing 25% of the total protein, followed by glutelins (7.3%), and prolamins (2%). All essential amino acids are present in pistachio with high lysine levels and a limited amount of cysteine. Pistachio nuts contain substantial levels of a diverse range of phytochemicals such as carotenoids (lutein), phytosterols, and phenolic compounds (flavonoid and resveratrol) in the kernel and hull (Bulló *et al.*, 2015; Barreca *et al.*, 2016; Ojeda-Amador *et al.*, 2018a, b). The presence of anthocyanins in pistachio nuts is a unique characteristic that causes the red-purple color in the hull. Anthocyanins are water-soluble pigments that impart the attractive red color.

Many studies have indicated that a diet containing pistachio nuts can reduce the risk of coronary heart disease (Bellocco *et al.*, 2016). Aside from their monounsaturated fatty acids and fiber, pistachio nuts are rich sources of antioxidant phytochemicals, which promote heart health by inhibiting the absorption of cholesterol from the intestine through direct competition with uptake mechanisms. Recent studies have shown that a diet that incorporates pistachio nuts can reduce the total cholesterol, total cholesterol/high-density lipoprotein ratio, and low-density lipoprotein/high-density lipoprotein ratio significantly, and decrease the plasma malondialdehyde, an important indicator of lipid peroxidation. Several studies have indicated that regular consumption of pistachio nuts does not lead to remarkable weight gain. Results have shown that those who ate pistachio nuts more frequently were leaner and had a lower body mass index than the infrequent eaters even though their energy intake was higher. It has also been proven that regular consumption of pistachio nuts lowers the blood pressure and therefore might be recommended for hypertension. They also reduce the absorption of glucose and lower the blood sugar. As a result of many vitamins and minerals, they are especially recommended for children for a healthy physical and mental development. Pistachio nuts are also a good source of vitamin E which boosts the immune system and alleviates fatigue. These days, it is strongly recommended to consume foods with minimal processing in order to gain maximum health benefits. Other than drying at low temperatures and sometimes roasting, no other treatment is commonly applied to pistachio. Therefore, fatty acids, minerals, vitamins, and other nutritional compounds are retained at maximum levels. Pistachios have been reported as a folk remedy for cirrhosis of the liver, abdominal ailments, abscess, amenorrhea, bruises, sores, trauma, and dysentery (Gentile *et al.*, 2015; Grace *et al.*, 2016).

Size and appearance are the most important components of pistachio quality. In the US Grades and Standards, quality factors include: (i) degree of dryness equivalent to 7.0% whole nut moisture content or lower; (ii) absence of foreign material and damage caused by mold or insects; and (iii) spotting or rancidity, and other external shell or kernel defects that can result in downgrading. External shell defects include total or partial non-splitting, adhering hull material, stained shells, and damage by other means, including deformity and bird damage. Internal kernel defects include damage from insects and fungal pathogens, small immature kernels, rancidity, and decay. In the USA, size, designated by the number of nuts per ounce, is also an important quality attribute. Grower payout is calculated from the fresh weight of pistachios delivered to the processing plant, corrected for the weight of foreign material removed prior to hulling. In-the-shell and shelled grades exist; these are determined primarily by kernel size, degree of dryness, absence of foreign material, and freedom from defects caused by insects and mold. For the in-the-shell product, additional grading criteria include absence of shell pieces and free kernels,

Fig. 5.2. Pistachio shell staining. Photo courtesy of Dr. Adel A. Kader, UC Davis, California, USA.

shells without stains and adhering hull material, and absence of non-split shells and blanks. A complete description of US quality standards, grades, and sizes for pistachios in the shell can be found at https://www.ams.usda. gov/grades-standards/pistachios-shell-grades-and-standards (accessed 11 May 2020).

Shell staining (Fig. 5.2) is usually caused by dehiscence of the hull along its suture at the same time as the shell within is splitting. This premature hull dehiscence increases "early" problems with insects and molds.

Fruit Physiological Characteristics

The pistachio tree is dioecious (meaning the male flowers are borne on one tree and the female flowers on another) with fruit-bearing perfect flowers on female trees being wind pollinated by imperfect, pollen-producing male-tree flowers. Female trees are smaller than male trees. Botanically, pistachio fruits (nuts) are semidry drupes, a classification they share with almonds, peaches, apricots, cherries, and plums. All drupes share a three-part structure: (i) the exocarp; (ii) the fleshy mesocarp; and (iii) the hard endocarp (Fig. 5.1). Together, the exocarp and mesocarp form the fleshy hull, while the endocarp is the shell that encloses the seed also known as the edible kernel (Polito and Pinney, 1999). As the oval-shaped nut of *P. vera* matures, its shell splits along a suture on one side while the hull remains at least partially intact. Upon harvest, the hull must be removed before it degrades and stains the shell or dries and adheres to the shell.

Respiration, ethylene production and sensitivity

Pistachio production of carbon dioxide and ethylene are very low. There are no documented responses to ethylene that might affect nut quality.

Chilling temperature sensitivity

Pistachios are not sensitive to chilling and can be stored at or below freezing.

Horticultural Maturity Indices

Physiological maturity is reached when the hull is easily separated from the shell, and the formation of the abscission zone between the nut and its rachis. Simultaneously, the kernel fat, sugar, and moisture contents, and respiration rate increase as the total protein content decreases and the kernel dry weight increases. Ideally, these changes usually coincide with splitting of the shell, however, shell split is not visible as the fleshy mesocarp masks the shell in developing nuts. Thus, nut maturity stage is determined when the hull color (exocarp) changes from green to ivory with blush development (Fig. 5.3). The percentage of blank and immature nuts decrease through maturation reaching their minimum at the optimum maturity. Harvest timing is very important in maintaining nut quality. Late harvests increase the potential for damage to the nut crop from: (i)

Fig. 5.3. Pistachios ready for harvest. Photo courtesy of Dr. Carlos H. Crisosto.

a fourth flight of navel orangeworm (NOW), *Amyelois transitella* (Walker); (ii) predation by birds; (iii) strong winds that remove nuts; (iv) early rains that halt harvest; and (v) an increased chance of fungal infestation (particularly *Aspergillus flavus*), and hull and kernel deterioration (Doster and Michailides, 1994). Furthermore, delayed harvest can lead to shell staining due to breakdown of the phenolics in the hull. Early harvest leads to weight loss and decreases the shelf life of fleshy harvested pistachios. Any hull deterioration present at harvest is aggravated during transport delays to the processor. At that stage, most of the nuts are split and moisture content ranges from 40% to 50%. Activity in the abscission zones between the nuts and the rachis (assessed by a measurable decrease in fruit removal force) also indicates maturation changes. A practical way to confirm if nuts are at the full maturity stage is by applying pressure with the thumb and forefinger at the hull's distal end; the mature pistachio shell will be ejected from the hull. Generally, the decision to start harvesting is taken when these maturity indexes are achieved in 70–80% of the nuts and harvesting equipment is available.

Physiological Disorders

There are several physiological disorders that decrease marketable yield. The first is alternate bearing, an annual fluctuation of large crops with small crops. The second is the production of "blanks": fully expanded shells containing either an unfertilized embryo or a fertilized embryo that aborted soon after fertilization or that failed to fully develop to maturity (Fig. 5.4). The third is non-split nuts (Fig. 5.5), nuts that contain a partially

Fig. 5.4. Pistachio blanks and non-blanks. Photo courtesy of Mr. Bob Beede, University of California Cooperative Extension (UCCE) Kings County, California, USA.

Fig. 5.5. Non-split pistachios. Photo courtesy of Dr. Louise Ferguson.

or fully developed kernel, but fail to split along the lateral shell suture. All three phenomena appear to be correlated with crop load and are therefore probably related to carbohydrate limitations or to embryo formation and development. Unfortunately, little is known about the specific physiological mechanism of each phenomenon, nor have control methods been developed. The last one is shell staining (Fig. 5.2).

Alternate bearing

Pistachios are alternate bearing in reference to the tendency of an entire tree to produce a greater than average crop one year, and a lower than average crop the following year. This cropping pattern is sometimes called on-year for a larger crop and off-year for a smaller crop and the failure of hull split and blank formation are correlated with "on" and "off" crop years. Studies indicate that blanks tend to be much higher in "off" years and non-split nuts are much more common in "on" years. Of course, crop load is much higher in "on" years, and this has a large effect on assimilate carbohydrate partitioning. Nut-bearing female pistachio trees display three physiological characteristics that directly affect marketable fruit yield though they occur at different stages of fruit development. Female pistachio trees bear their nuts laterally on 1-year-old shoots. Once female trees are mature, virtually all the axillary lateral buds on the current year's shoot growth are inflorescence buds that will grow and develop into next year's crop. Remember that inflorescence buds produce flowers that generate pistachio nuts. The single apical bud beyond the flower buds will produce the following year's shoot growth and contains the preformed flower buds to be produced the following year. Therefore, the individual bearing shoots within a tree are producing the current year's crop as well as generating the growth and buds for the subsequent year's crop. Producing a crop while simultaneously generating shoots with reproductive buds appears to be

the visual mechanism of alternate-bearing common to all pistachio cultivars. The simultaneous demand of maturing the current crop on 1-year-old
shoots while producing the shoot growth that bears the inflorescence buds
for the subsequent crop exerts a high carbohydrate demand on the tree's
available resources. This visible resource demand is easily observed during
the growing season on shoots bearing a heavy load of nuts. The inflorescence buds on the current year's shoot growth beyond the nuts abscise by
mid-August. There is some evidence that the shoots cannot support a maturing crop and develop reproductive buds. Ultimately the tree drops the
reproductive buds as the current crop matures. As the dropped buds were
to be next year's crop, the following year's crop on that shoot is noticeably
smaller.

Blank nuts

Blank nuts (Fig. 5.4) result when the embryo fails to develop during shell
development from ovary tissues without successful fertilization leaving the
shell empty, or blank. Blanking can occur during two different phases of
pistachio nut development: (i) pollination and nut set; and (ii) nut fill
(Ferguson and Polito, 2016) The first mechanism of blank nut production
before nut set is parthenocarpy; the shell is stimulated to grow with pollination, but without the fertilization that produces an embryo, thus, the
nutshell remains empty. The second mechanism is if fertilization is successful a nut is set, however, this young embryo may abort after set, producing
a blank nut through embryo abortion. The third mechanism that produces
blank nuts occurs during nut fill when a fertilized and set embryo fails to
enlarge, or fill, again creating a blank. As this blank development occurs
later in the season, it has been proposed as the result of the inability of
the tree to provide enough assimilate to complete development of its entire crop. Unfortunately, the mechanism triggering nut blanking or how
the processes interact within the three is not well known. Developmental
and physiological problems that occur before full maturity can have particularly important consequences for postharvest nut quality. Because nuts
are only useful when they have split, failure of hull split as nuts reach full
maturity can cause substantial yield losses. Inadequate boron and water
stress are also indicated as causes of blank formation. While splitting is
maturation-dependent, it will be reduced by water stress late in the growing season (mid-August through to September) and failure to maintain
adequate boron (120 mg kg^{-1} leaf dry weight).

Non-split nuts

Commercially grown pistachio is characterized by splitting of the nutshell
at the beginning of maturity (Fig. 5.6). A non-split nut (Fig. 5.5) is not fully
mature and lacks the final dehiscence, or longitudinal split, of the shell

Fig. 5.6. Pistachio shell split. Photo courtesy of Dr. Louise Ferguson.

(Freeman and Ferguson, 1995). In California, splitting begins about the end of July, at least 1 month before fruit maturity, and continues through to mid-September, progressing simultaneously with nutmeat maturation (Ferguson and Polito, 2016). It has been demonstrated that shell split is dependent on nutmeat growth and development within the shell. By June of the growing season, shell size has already reached its maximum, or full size, when kernel growth begins. The first split shells are not seen until late July when the kernel has grown to fill the fully sized shell and begins exerting physical pressure on the shell. At this time, the shell is fully lignified or hardened, and the cells that form the regions where longitudinal splitting occurs are dead, seemingly ruling out the possibility that a biochemical mechanism controls shell splitting. The lack of a biochemical factor makes it unlikely a chemical agent will be discovered to enhance splitting.

Fortunately, three factors can be managed to maximize shell split: (i) harvest timing; (ii) irrigation; and (iii) boron nutrition. To obtain the best split shell percentages with the 'Kerman' cultivar: (i) harvest timing should be based on the appearance of hull color and a sampling for the split percentage; (ii) trees should not be water stressed from mid-August through to September; and (iii) boron levels should be maintained above 120 mg kg^{-1} by dry weight of a July leaf sample.

Shell staining

This is one of the biggest concerns during postharvest transport and storage, and can largely increase if high levels of hull damage were sustained during harvest. Shell staining generally increases with increased temperatures and increased holding times (delays). Even at night, the temperature of nuts held in trailers increases up to ~1.0°C h^{-1}. Nuts with good-quality, intact hulls may be held for up to 48 h at ambient conditions without an increase in staining. Nuts with poor-quality hulls show damage after only 8 h at 40°C, 24 h at 30°C and 40 h at 25°C.

Harvesting and Postharvest Operations

As soon as separation of the shell is observed (early fall), one important factor in harvest preparation is to monitor irrigation, as pistachio shell splitting is particularly sensitive to irrigation deficits that decrease the number of split nuts and pistachio quality (Ferguson *et al.*, 1995; Thompson and Kader, 2016) (Fig. 5.7). Therefore, irrigation should be carefully monitored, preferably using budget irrigation methods, through the August and early September shell-splitting period. The approach is to maintain adequate soil moisture to maximize shell splitting while ensuring that the orchard rows will be dry enough to support harvest equipment. The second factor in harvest preparation is monitoring for NOW. Infestation of NOW results in unsightly damage by the larvae and increases the susceptibility of the nut to infection with *Aspergillus* spp., a fungus that produces a carcinogenic aflatoxin (Doster and Michailides, 1994, 1999). Depending on temperatures, calculated as degree days, the third generation of NOW emerges in early to mid-August and lays eggs in the sutures of the early split nuts where the shells split before the hulls dehisce or split. When the shell splits prematurely, the hull also splits and exposes the kernel to NOW infestation. Monitoring for NOW is to assess the quantity of early split nuts. If early splits are higher than one or more nuts per cluster, the early split nuts should be monitored for eggs. If egg laying is present, the trees should be sprayed with an insecticide to reduce damage to the nuts. Early split nuts are usually no more than 1–5% of a tree's total nut load, but an infestation

Fig. 5.7. Pistachio orchard prepared for harvest. Photo courtesy of Dr. Carlos H. Crisosto.

of even 1% of the nuts can result in aflatoxin levels above maximum allowable levels, 20 µg kg⁻¹ in the USA and 4 µg kg⁻¹ in the European Union.

In most countries, trees are shaken (by hand for young trees, mechanically for mature trees) and the nuts are caught on tarps or a catching frame and transferred to bins in order to eliminate problems caused by contact with the soil. Minimizing delays between harvest and further processing will reduce potential hull breakdown and contamination (staining) of hull tissues. Deterioration problems including staining triggered by unavoidable delays in hulling and drying (below) can be reduced by cold storage of bins at 0°C and <70% RH without increasing shell staining. At the processor, the bins of nuts are dumped, and debris is removed by passage over an air leg. Hulls are removed, blanks are removed in a float tank, and the in-shell nuts are dried to 5–7% moisture. Most large handlers now use a two-stage process: nuts are dried in a column dryer to 12–13% moisture with forced hot air at 82°C before the drying is completed more slowly (24–48 h) with air heated to no more than 49°C.

Shaking harvesting

The principles are the same for both young and full-bearing pistachio trees. The nuts are removed by knocking or shaking on to a catching frame. Because of the high moisture content of 40–50% on a fresh-weight basis, fragility of their ripe hulls, and open shells within the hull, pistachios are susceptible to mechanical injury, shell staining, and contamination if they drop to the orchard floor. As *A. flavus* and other *Aspergillus* spp. are present in wet orchard soils and have the potential to infect pistachios that contact the soil, it is very important to prevent dropping the pistachios to the ground during harvesting. In the USA, young trees are hand harvested by knocking the trunk or striking the branches to knock the nuts on to tarps spread under the trees, while mature pistachios are mechanically harvested on to a catching frame. For harvesting young pistachio trees, tarps are spread under both sides of the tree to cover at least 1.5 m beyond the edge of the canopy on all sides. Nuts are removed by knocking the tree trunk with a padded mallet or striking the branch behind the clusters with a pole. When nuts cease dropping, the tarps are gathered by the edges and lifted vertically. This brings most of the leaves and lightweight debris to the surface for removal prior to dumping into wood or plastic field bins that are 1.2 m × 1.2 m × 0.6 m. Bins hold about 454 kg of freshly harvested nuts and should have at least 5% of the vertical surfaces occupied by air vents. These bins are distributed in the orchard prior to harvest, filled with hand-harvested nuts, and then transported to the edge of the orchard by mechanical bin carriers. Forklifts are used to load the bins on to trailers for transport to the processor. The bins can also be dumped into hopper-bottomed trailers and the nuts hauled in bulk to the processor. On mature pistachio trees, unless the crop is too light to justify the cost of machine harvesting which can occur on light-crop years of this alternate-bearing crop, nuts are mechanically harvested. Also, in orchards where space between trees and rows do not allow the use of mechanical harvesting equipment (as occurs in some locations in Iran, Turkey, and Syria) pistachio harvesting is carried out by hand in clusters and dumped into sacks. The sacks are collected into plastic bins and transported to processing plants by trucks or trailers.

Mechanical harvesting

A pistachio harvester that consists of two separate, self-propelled units about 7.3 m in length is used (Fig. 5.8). One unit contains a shaker head that is clamped on to the tree trunk about 0.6 m above the ground. This unit matches with the other unit to form a continuous collection surface under the tree canopy. Both units have gently sloping wings extending under the canopies into the row middles. After the tree has been shaken for 6–8 s, the wings direct the nuts into the bottom of one catch frame. The nuts are then conveyed over an air separator to remove light debris. The major difference in harvester systems is in how the nuts are handled after shaking. In a bin system, the nuts are collected in pallet bins that have been distributed in the orchard before harvest (Fig. 5.9). After filling, they are moved by mechanical bin carriers to the edge of the orchard for collection. The bins are either stacked on flatbed trucks or dumped into

Fig. 5.8. Shake-and-catch harvesting system on a mature tree. Photo courtesy of Dr. Carlos H. Crisosto.

hopper-bottomed trailers for transport to the processor (Fig. 5.10). The bulk system consists of carts carried behind the shaker receiver. The carts often have large, central, continuously revolving screws to distribute the crop in the machine. When filled, the carts are unloaded by a continuous transport system or left at the end of the row to be unloaded into bulk trailers by an elevator system. Larger operations generally use a bulk system, as it is more efficient for large volumes. Most harvesters have the capacity to cover 1 acre h^{-1} (1 acre contains 112 female trees).

Orchard to huller

In California, shell staining is the biggest concern during postharvest transport and storage as it can largely increase if high levels of hull damage were sustained during harvest. Shell staining generally increases with increased temperatures and increased holding times (delays). Even at night, the temperature of nuts held in trailers increases up to ~1.0°C h^{-1}. Nuts with good-quality, intact hulls may be held for up to 48 h at ambient conditions without an increase in staining (Thompson *et al.*, 1997). Nuts with poor-quality hulls show damage after only 8 h at 40°C, 24 h at 30°C and 40 h at 25°C. Nuts transported in bins, leaving the orchard at the same time as the bulk trailers, will arrive cooler. Therefore, if travel time between the orchard and the processor is particularly long, or delays at the processor are anticipated, plastic bins with at least 5% of the vertical surfaces vented are

Fig. 5.9. Filling bins after shaking. Photo courtesy of Dr. Carlos H. Crisosto.

the preferred mode of transport. Bulk trailers with mesh sides are preferable to solid trailers. Thus, if long delays are anticipated prior to transport or at the processor, keeping the bins or hoppers in the shade will provide some protection from the elevated temperatures that cause shell staining.

Hulling process

When pistachios arrive at the processing plant, they move through a series of procedures. Pistachios delivered in tared flatbed or bulk trailers are weighed and tagged for delivery fresh weight. Temperature within the load is measured and nuts are dumped and conveyed over an air leg to remove debris. A 9.1 kg unhulled sample is collected for separate processing

Fig. 5.10. Hauling pistachios to the processing plant. Photo courtesy of Dr. Carlos H. Crisosto.

and grading. If the hulling capacity is limited, priority should be given to nuts with the highest internal nut temperatures and poorest hull quality. If delays over 4 h occur at the huller, circulating ambient air through bins will prevent temperature increases and nut deterioration for up to 2 days. Hauling trailers at highway speeds induces air ventilation in the unhulled nuts and cools them. If the anticipated delay is longer than 2 days, keeping the bins in cold storage at 0°C and less than 70% RH is an alternative. If possible, the pistachios should be sorted before storage to remove debris and damaged nuts that are more susceptible to decay, as it will minimize storage losses (Tavakolipour, 2015; Nazoori *et al.*, 2017). Sound unhulled pistachios can be stored for up to 6 weeks at 0°C and 70–75% RH without any significant effects on appearance, flavor quality, composition, and hull removal. Storage at higher temperatures results in high incidence of surface molds and shell staining. An adequate air flow rate of 0.104 m^3 s^{-1} kg^{-1} through nuts during storage is essential to minimize losses. Hulled pistachio nuts have a shorter shelf life because the hull protects the nut from decay organisms without affecting shell quality. Hulled fresh pistachio nuts can be held for up to 3 weeks at the best storage conditions of 0°C and 40–50% RH. Increasing the RH to 90% shortens the storage of hulled pistachio nuts to 2 weeks at 0°C, 1 week at 5°C, 4 days at 10°C, 2 days at 20°C and 1 day at 30°C. After 2 weeks storage at 0°C and less than 10% RH the moisture content decreases to 7% and hulled nuts are almost completely dry. Although freezing has no significant effects on kernel texture, its influence on flavor quality eliminates it as an alternative for extending shelf life of fresh hulled nuts.

Prior to hulling, a grading sample is drawn from each delivery load. This sample is processed individually, the fresh-to-dried-weight ratio is calculated, and a third party inspects the sample to assess the percentage by weight of split nuts, non-split nuts, blank nuts, nuts with adhering hulls, light and dark stain, and other defects. Correction factors for these defects, along with the cost of hulling and drying, are applied to the corrected delivery weight to calculate the grower price. The major components in determining return to the grower are the weight delivered and the percentages by weight of the filled split, filled non-split, and blank nuts. Aflatoxin contamination is associated with nut defects, which include: (i) nuts that freely separate from their shell; (ii) nuts that are smaller than 40 individual nuts per ounce; (iii) severe shell staining; (iv) evidence of insect damage; and (v) large nuts that float during density separation prior to drying. Removing these nuts can reduce contamination in a lot.

Historically in old pistachio production, pistachio nuts were processed manually. After harvesting the ripe pistachios were hulled by hand and in other regions they were dried in the hull for later hulling at a convenient time. Sometimes the pistachio nuts were immersed in water to separate the hull easily when squeezed between the fingers. These days, hulling must be accomplished as soon as possible to reduce the chances for fungal growth and to avoid shell staining. The hulls of the freshly harvested and mature pistachio nuts slip off fairly easily if the hull has not begun to shrivel. Generally, different machines used are satisfactory for hulling pistachios. The cylindrical hullers are popular and different sizes and capacities are used to hull pistachios in various regions of Iran and Turkey, whereas hulls are removed from nuts with an abrasive peeler in California.

Processing sorting

Blank nuts are separated by float tank and separately dried and stored. Filled split and non-split nuts are dried to 12–13% moisture in a high-temperature dryer (first drying step). The nuts are stored at ambient temperature with forced air ventilation. Split nuts are separated from non-splits by a needle-picking drum. Non-splits are shelled or split mechanically. Split nuts are sorted by an electronic color reflectance sorter and split nuts are hand-graded to remove defects and debris and stored or shipped (Pearson, 1996; Pearson and Slaughter, 1996; Slaughter *et al.*, 1996).

Drying operations

Currently, drying of pistachios is generally a two-stage process. It uses less energy and increases the output of the heated air dryer compared with the single-stage process. In the first stage, the hulled nuts are dried to 12–13%

Fig. 5.11. Pistachio first-stage dryer. Photo courtesy of UC ANR, Davis, California, USA.

moisture content in a column dryer, originally designed for grain drying (Fig. 5.11), or a continuous belt dryer. In these driers, the nuts dry in about 3 h at temperatures near 82.2°C. Drying at air temperatures above 82.2°C causes shells to split open so widely that the kernel drops out. A rotating drum dryer, at the same temperature, can also be used for the first stage of drying. In the second stage the nuts are transferred to grain bins equipped with perforated floors (Fig. 5.12), where they are further dried to 4–6% moisture with unheated, forced air, or air heated to less than 48.9°C. Weather in California during harvest is usually warm and dry and the second stage of drying requires only 24–48 h. The nuts can then be stored in these bins until they are needed for processing. Smaller operations may use

Fig. 5.12. Pistachio second-stage dryer and storage. Photo courtesy of UC ANR, Davis, California, USA.

bin dryers for single-stage drying; 8 h at 60–65.6°C produces the desired 4–6% moisture content. In very small operations, sun drying, and ambient air drying may also be used. Drying in the sun requires 3–4 days, with a protective covering to prevent predation by birds and rodents. Drying nuts in bins with ambient air requires 3 days in California. Nut depth in the bin should not exceed 1.4 m, and air should flow into the bins at a rate of 0.0078 m^3 s^{-1} per 0.00283 m^3 of nuts. The major disadvantage to this method is the potential for fungal growth during the early part of drying. Once dried to 4–6% moisture content, but before further processing, nuts can be held at 20°C and 65–70% RH for up to 1 year. Lower temperatures slow lipid oxidation (which results in rancidity), prevents mold growth, and greatly reduces insect activity. Temperatures between 0°C and 10°C are recommended for long-term storage of pistachios, depending on expected storage duration, the lower the temperature, the longer the storage life.

Temperature Handling, Optimum Storage and Shipping Conditions

Once they have been dried, nuts can be held at 20°C and 65–70% RH for up to a year. Pistachios are considerably less prone to rancidification (precipitated by oxidation of polyunsaturated fatty acids) than are almonds and, particularly, pecans and walnuts. These commodities are also high in fat content, but walnut and pecan oils have a much higher content of polyunsaturated fatty acids than pistachio oil. The US Food and Drug Administration defines "safe moisture levels" in terms of water activity according to a given temperature and RH. Water activity is measured by a device and is expressed as decimal units from 0–1.0. For example, water activity levels less than 0.70 at 25°C do not support growth of fungal and human pathogens on whole nuts with a moisture content of approximately 7.0%. Other factors that reduce pistachio quality deterioration during storage are low temperatures, exclusion of oxygen through controlled atmosphere storage or packaging under nitrogen, and fumigation or insect-proof packaging to prevent insect damage (Ozturk *et al.*, 2016). Generally, nuts may be held for a long time, up to 2–5 years, if they are stored under optimum conditions, but with unfavorable storage conditions they can become inedible even within 1 month, because of mold growth, off-flavor, rancidity, discoloration, absorption of undesirable flavors and insect infestation (Tavakoli *et al.*, 2019). Pistachio nuts dried to the appropriate moisture content (4–6%) are very stable and can be stored for up to 1 year at 20°C and 65–70% RH without significant losses in quality attributes. No differences in chemical composition and sensory attributes were observed among nuts stored at different temperatures (0°C, 5°C, 10°C, 20°C and 30°C) for 12 months. Nuts held at 30°C had a lower moisture content and higher sugar content and rancid flavor than those stored at lower temperatures.

Controlled atmosphere storage

For longer storage of pistachio nuts temperatures between 0°C and 10°C are recommended. Exclusion of oxygen, insect control through fumigation, vacuum packaging or nitrogen injection in packages and controlled atmospheres can maintain quality during storage. Storage under a high concentration of carbon dioxide (98%) and reduced oxygen (less than 0.5%) provides good stability in terms of fatty acid loss and formation of peroxide and free fatty acids (rancidity precursors). Oxygen scavenger packaging is an efficient method to eliminate the oxygen content and retard the oxidation of pistachio nuts during long storage (Maskan and Karataş, 1998).

Modified atmosphere packaging (MAP)

The postharvest quality of hulled fresh 'Kerman' pistachios was protected using a passive MAP for up to 105 days when stored at 0°C compared with 30 days in ambient conditions. A modified atmosphere that fluctuated from 0.9–3.3% O_2 and 23–29% CO_2 was achieved in the passive-MAP treatment that maintained firmness, shell lightness, kernel color, and sensory quality with minimum weight loss and fungal decay, compared with the control (Sheikhi *et al.*, 2019a). Further work comparing active MAP to passive MAP concluded that active MAP was more effective than the passive MAP in decreasing weight loss, microbial counts, kernel total chlorophyll, and kernel carotenoid content. The postharvest quality of fresh in-hull pistachios was maintained best by the active MAP at 5% O_2 + 45% CO_2 balanced with N_2 (Sheikhi *et al.*, 2019b).

Postharvest Pathology

Several fungi are capable of infecting growing pistachio nuts and causing damage to hulls and kernels. Infection is often facilitated by early splitting of hulls, which leads to infestations by several hemipteran insects that feed on the nuts and serve as non-specific vectors for diseases (Michailides *et al.*, 2016). *Alternaria* and *Cladosporium* species are also colonizers of early split nuts. Late season rains will promote activity of *Botryosphaeria dothidea* on pistachio hulls and kernels. Because mold counts on nuts going into storage can be high, it is important that proper storage conditions (especially low RH and absence of standing water) be maintained to avoid serious problems (Aldars-García *et al.*, 2015, 2016). The greatest postharvest disease threats are from *A. flavus* and *Aspergillus parasiticus*. The danger is particularly serious because these fungi can produce aflatoxin (Steiner *et al.*, 1992; Heperkan *et al.*, 1994). As with many disease problems of pistachio, vectoring by insects attracted to early split nuts (such as NOW) is an important contributing factor.

Kernel Utilization

Pistachios are mainly eaten as snacks: fresh, dried and roasted, with or without salt and flavorings. The pistachio is different from other nuts as the shell splits naturally before harvesting allowing it to be marketed in-shell for fresh consumption. Also, as kernels can be easily separated without mechanical cracking, this enables processors to roast and salt the kernel without shell removal as is done with almonds, walnuts and pecans that are generally sold shelled (Ghazzawi and Al-Ismail, 2017; Bai *et al.*, 2019). Pistachios are used in pastries, cakes, ice creams, confections, baked goods, candies, sausages and desserts. Pistachio is an excellent taste enhancer and can be added to many food products to improve nutrition, color, and flavor. A small amount of pistachio oil is produced for eating and use in cosmetics, while a limited amount of pistachio is also used for the production of pistachio butter (Kilic *et al.*, 2016). This butter, a semi-solid paste, is made from ground and roasted pistachio kernels with addition of some flavorings and sweeteners. Pistachio butter is a nutritive product rich in lipids, proteins, carbohydrates, and vitamins and can be used in different food products such as flour (Martínez *et al.*, 2016), cookies, ice cream, and cakes.

Cull Utilization (Hulls and Shells)

Limited research indicates that cull pistachios, which typically represent 1.5–2.5% of the marketable nuts, can be fed to sheep and cattle. However, the sharp edges of broken nutshells can harm animal intestines. Rations for cattle can contain up to 20% culls and for sheep up to 10%. Pistachio

Fig. 5.13. Pistachio warm retail display. Photo courtesy of Dr. Pat J. Brown, UC Davis, California, USA.

hulls have high tannin and phenolic contents that limit their use in animal feed. Some processing plants give the hulls to cogeneration plants that generate electricity and heat for free (Ahanchi *et al.*, 2018). The hull is used as a fertilizer and ruminant feed, an anti-browning agent (Fattahifar *et al.*, 2018), for dyeing and tanning in India, and small amounts are made into a flavorful green hull marmalade.

Retail Outlet Display Considerations

Dried pistachios are packed into various types of film bag and placed in different large containers such as promotional bins, mini pallets, floor displays, and shelf-insert display cases. All these containers, mainly of corrugated material, have very colorful graphics and information on the health benefits of pistachios to attract consumers to the display. Pistachios are sold as in-shell roasted and salted, lightly salted, unsalted, with salt and pepper, or sweet chili, or as raw shelled, and these are packed in different sized packages that vary from 227 g to 907 g. These packages are placed in different sized containers and in all cases the pistachios are displayed in large exhibits at warm temperatures to attract consumers (Fig. 5.13).

Quarantine Issues

Insect infestation is a potentially important problem, as are the fungal infections that often accompany insect damage. Several insects that are field pests of pistachios can cause superficial damage ("epicarp lesion") to developing nuts. If insects can probe deeply or introduce fungal pathogens, these pests can cause damage to the kernels. The NOW, *A. transitella* (Walker), a primary field pest, is the major insect problem after harvest. Fumigation with methyl bromide or phosphine has been used for disinfestation (Hartsell *et al.*, 1986), but the former is being curtailed and insect resistance to phosphine has been reported (Zettler *et al.*, 1989). New fumigants are being developed and tests of efficacy, including effects on flavor, are being performed. New fumigants are not registered for pistachio nuts.

Special Research Needs

The research emphasis should be as follows:

- pursue commercial conversion technologies that transform pistachio residues to energy and other valuable products;
- continue searching for phytochemicals and antioxidant activities of pistachio hulls; and
- study the potential value of by-products such as biochar for soil amendment and bioliquids for biopesticide applications.

References

Ahanchi, M., Tabatabaei, M., Aghbashlo, M., Rezaei, K., Talebi, A.F. *et al.* (2018) Pistachio (*Pistachia vera*) wastes valorization: enhancement of biodiesel oxidation stability using hull extracts of different varieties. *Journal of Cleaner Production* 185, 852–859. DOI: 10.1016/j.jclepro.2018.03.089.

Aldars-García, L., Ramos, A.J., Sanchis, V. and Marín, S. (2015) An attempt to model the probability of growth and aflatoxin β1 production of *Aspergillus flavus* under non-isothermal conditions in pistachio nuts. *Food Microbiology* 51, 117–129. DOI: 10.1016/j.fm.2015.05.013.

Aldars-García, L., Ramos, A.J., Sanchis, V. and Marín, S. (2016) Modelling the probability of growth and aflatoxin B1 production of *Aspergillus flavus* under changing temperature conditions in pistachio nuts. *Procedia Food Science* 7, 76–79. DOI: 10.1016/j.profoo.2016.02.091.

Bai, S.H., Brooks, P., Gama, R., Nevenimo, T., Hannet, G. *et al.* (2019) Nutritional quality of almond, canarium, cashew and pistachio and their oil photooxidative stability. *Journal of Food Science and Technology* 56(2), 792–798. DOI: 10.1007/s13197-018-3539-6.

Barla Demirkoz, A., Karakaş, M., Bayramoğlu, P. and Üner, M. (2018) Analysis of volatile flavour components by dynamic headspace analysis/gas chromatography-mass spectrometry in roasted pistachio extracts using supercritical carbon dioxide extraction and sensory analysis. *International Journal of Food Properties* 21(1), 973–982. DOI: 10.1080/10942912.2018.1466322.

Barreca, D., Laganà, G., Leuzzi, U., Smeriglio, A., Trombetta, D. *et al.* (2016) Evaluation of the nutraceutical, antioxidant and cytoprotective properties of ripe pistachio (*Pistacia vera* L., variety Bronte) hulls. *Food Chemistry* 196, 493–502. DOI: 10.1016/j.foodchem.2015.09.077.

Bellocco, E., Barreca, D., Laganà, G., Calderaro, A., El Lekhlifi, Z. *et al.* (2016) Cyanidin-3-O-galactoside in ripe pistachio (*Pistachia vera* L. variety Bronte) hulls: identification and evaluation of its antioxidant and cytoprotective activities. *Journal of Functional Foods* 27, 376–385. DOI: 10.1016/j.jff.2016.09.016.

Bulló, M., Juanola-Falgarona, M., Hernández-Alonso, P. and Salas-Salvadó, J. (2015) Nutrition attributes and health effects of pistachio nuts. *British Journal of Nutrition* 113 Suppl. 2, S79–S93. DOI: 10.1017/S0007114514003250.

Carughi, A., Bellisle, F., Dougkas, A., Giboreau, A., Feeney, M.J. *et al.* (2019) A randomized controlled pilot study to assess effects of a daily pistachio (*Pistacia vera*) afternoon snack on next-meal energy intake, satiety, and anthropometry in French women. *Nutrients* 11(4), E767. DOI: 10.3390/nu11040767.

Doster, M.A. and Michailides, T.J. (1994) *Aspergillus* molds and aflatoxins in pistachio nuts in California. *Phytopathology* 84(6), 583–590. DOI: 10.1094/Phyto-84-583.

Doster, M.A. and Michailides, T.J. (1999) Relationship between shell discoloration of pistachio nuts and incidence of fungal decay and insect infestation. *Plant Disease* 83(3), 259–264. DOI: 10.1094/PDIS.1999.83.3.259.

Fattahifar, E., Barzegar, M., Ahmadi Gavlighi, H. and Sahari, M.A. (2018) Evaluation of the inhibitory effect of pistachio (*Pistacia vera* L.) green hull aqueous extract on mushroom tyrosinase activity and its application as a button mushroom postharvest anti-browning agent. *Postharvest Biology and Technology* 145, 157–165. DOI: 10.1016/j.postharvbio.2018.07.005.

FAOSTAT (2018) Pistachios. Available at: http://www.fao.org/faostat/en/#search/pistachio (accessed 11 May 2020).

Ferguson, L. and Kallsen, C.E. (2016) The pistachio tree: physiology and botany. In: Ferguson, L. and Haviland, D.R. (eds) *Pistachio Production Manual.* UCANR Publications, Richmond, California, pp. 19–26.

Ferguson, L.F. and Polito, V.S. (2016) Alternate bearing, nut blanking, and shell splitting. In: Ferguson, L. and Haviland, D.R. (eds) *Pistachio Production Manual.* UCANR Publications, Richmond, California, pp. 313–321.

Ferguson, L., Kader, A. and Thompson, J. (1995) Harvesting, transporting, processing and grading. In: Ferguson, L. (ed.) *Pistachio Production.* University of California, Pomology Department, Center for Fruit and Nut Crop Research and Information, Davis, California, pp. 110–114.

Freeman, M. and Ferguson, L. (1995) Factors affecting splitting and blanking. In: Ferguson, L. (ed.) *Pistachio Production.* University of California, Pomology Department, Center for Fruit and Nut Crop Research and Information, Davis, California, pp. 106–109.

Gentile, C., Perrone, A., Attanzio, A., Tesoriere, L. and Livrea, M.A. (2015) Sicilian pistachio (*Pistacia vera* L.) nut inhibits expression and release of inflammatory mediators and reverts the increase of paracellular permeability in IL-1β-exposed human intestinal epithelial cells. *European Journal of Nutrition* 54(5), 811–821. DOI: 10.1007/s00394-014-0760-6.

Ghasemi-Varnamkhasti, M. (2015) Sensory stability of pistachio nut (*Pistacia vera* L.) varieties during storage using descriptive analysis combined with chemometrics. *Engineering in Agriculture, Environment and Food* 8(2), 106–113. DOI: 10.1016/j.eaef.2014.11.002.

Ghazzawi, H.A. and Al-Ismail, K. (2017) A comprehensive study on the effect of roasting and frying on fatty acids profiles and antioxidant capacity of almonds, pine, cashew, and pistachio. *Journal of Food Quality* 2017(1): 9038257. DOI: 10.1155/2017/9038257.

Grace, M.H., Esposito, D., Timmers, M.A., Xiong, J., Yousef, G. *et al.* (2016) Chemical composition, antioxidant and anti-inflammatory properties of pistachio hull extracts. *Food Chemistry* 210, 85–95. DOI: 10.1016/j.foodchem.2016.04.088.

Hartsell, P.L., Nelson, H.D., Tebbets, J.C. and Vail, P.V. (1986) Methyl bromide fumigation treatments for pistachio nuts to decrease residues and control navel orangeworm, *Amyelois transitella* (Lepidoptera: Pyralidae). *Journal of Economic Entomology* 79(5), 1299–1302. DOI: 10.1093/jee/79.5.1299.

Heperkan, D., Aran, N. and Ayfer, M. (1994) Mycoflora and aflatoxin contamination in shelled pistachio nuts. *Journal of the Science of Food and Agriculture* 66(3), 273–278. DOI: 10.1002/jsfa.2740660302.

Kader, A.A., Heintz, C.M., Labavitch, J.M. and Rae, H.L. (1982) Studies related to the description and evaluation of pistachio nut quality. *Journal of the American Society for Horticultural Science* 107, 812–816.

Kashaninejad, M. and Tabil, L.G. (2011) Pistachio (*Pistacia vera*. L.). In: Yahia, E.M. (ed.) *Postharvest Biology and Technology of Tropical and Subtropical Fruits. Volume 4: Mangosteen to White Sapote.* Woodhead Publishing Limited, Cambridge, UK, pp. 218–246.

Kilic, I.H., Sarikurkcu, C., Karagoz, I.D., Uren, M.C., Kocak, M.S. *et al.* (2016) A significant by-product of the industrial processing of pistachios: shell skin – RP-HPLC analysis, and antioxidant and enzyme inhibitory activities of the methanol extracts of *Pistacia vera* L. shell skins cultivated in Gaziantep, Turkey. *RSC Advances* 6(2), 1203–1209. DOI: 10.1039/C5RA24530C.

Labavitch, J.M., Heintz, C.M., Rae, H.L. and Kader, A.A. (1982) Physiological and compositional changes associated with maturation of 'Kerman' pistachio nuts. *Journal of the American Society for Horticultural Science* 107, 688–692.

Martínez, M.L., Fabani, M.P., Baroni, M.V., Huaman, R.N.M., Ighani, M. *et al.* (2016) Argentinian pistachio oil and flour: a potential novel approach of pistachio nut utilization. *Journal of Food Science and Technology* 53(5), 2260–2269. DOI: 10.1007/s13197-016-2184-1.

Maskan, M. and Karataş, Ş. (1998) Fatty acid oxidation of pistachio nuts stored under various atmospheric conditions and different temperatures. *Journal of the Science of Food and Agriculture* 77(3), 334–340. DOI: 10.1002/(SICI)1097-0010(199807)77:3<334::AID-JSFA42>3.0.CO;2-A.

Michailides, T., Morgan, D.P. and Doster, M.A. (2016) Foliar, fruit, and branch diseases. In: Ferguson, L. and Haviland, D.R. (eds) *Pistachio Production Manual.* UCANR Publications, Richmond, California, pp. 265–292.

Nazoori, F., Kalantari, S., Javanshah, A. and Talaie, A.R. (2017) The effect of different storage conditions on quantitative and qualitative properties Ahmad Aghaie fresh pistachio. *Iranian Journal of Food Science and Technology* 14, 65–74.

Ojeda-Amador, R.M., Fregapane, G. and Salvador, M.D. (2018a) Composition and properties of virgin pistachio oils and their by-products from different cultivars. *Food Chemistry* 240, 123–130. DOI: 10.1016/j.foodchem.2017.07.087.

Ojeda-Amador, R.M., Trapani, S., Fregapane, G. and Salvador, M.D. (2018b) Phenolics, tocopherols, and volatiles changes during virgin pistachio oil processing under different technological conditions. *European Journal of Lipid Science and Technology* 120(10), 1800221. DOI: 10.1002/ejlt.201800221.

Ozturk, I., Sagdic, O., Yalcin, H., Capar, T.D. and Asyali, M.H. (2016) The effects of packaging type on the quality characteristics of fresh raw pistachios (*Pistacia vera* L.) during the storage. *LWT - Food Science and Technology* 65, 457–463. DOI: 10.1016/j.lwt.2015.08.046.

Pantano, L., Lo Cascio, G., Alongi, A., Cammilleri, G., Vella, A. *et al.* (2016) Fatty acids determination in Bronte pistachios by gas chromatographic method. *Natural Product Research* 30(20), 2378–2382. DOI: 10.1080/14786419.2016.1180599.

Parfitt, D.E., Kallsen, C.E. and Maranto, J. (2016) Pistachio cultivars. In: Ferguson, L. and Haviland, D.R. (eds) *Pistachio Production Manual.* UCANR Publications, Richmond, California, pp. 59–64.

Pearson, T. (1996) Machine vision system for automated detection of stained pistachio nuts. *LWT - Food Science and Technology* 29(3), 203–209. DOI: 10.1006/fstl.1996.0030.

Pearson, T.C. and Slaughter, D.C. (1996) Machine vision detection of early split pistachio nuts. *Transactions of the ASAE* 39(3), 1203–1207. DOI: 10.13031/2013.27613.

Polari, J.J., Zhang, L., Ferguson, L., Maness, N.O. and Wang, S.C. (2019) Impact of microclimate on fatty acids and volatile terpenes in 'Kerman' and 'Golden Hills' pistachio (*Pistacia vera*) kernels. *Journal of Food Science* 84(7), 1937–1942. DOI: 10.1111/1750-3841.14654.

Polito, V.S. and Pinney, K. (1999) Endocarp dehiscence in pistachio (*Pistacia vera* L.). *International Journal of Plant Sciences* 160(5), 827–835. DOI: 10.1086/314186.

Rodríguez-Bencomo, J.J., Kelebek, H., Sonmezdag, A.S., Rodríguez-Alcalá, L.M., Fontecha, J. *et al.* (2015) Characterization of the aroma-active, phenolic, and lipid profiles of the pistachio (Pistacia vera L.) nut as affected by the single and double roasting process. *Journal of Agricultural and Food Chemistry* 63(35), 7830–7839. DOI: 10.1021/acs.jafc.5b02576.

Sheikhi, A., Mirdehghan, S.H. and Ferguson, L. (2019a) Extending storage potential of de-hulled fresh pistachios in passive-modified atmosphere. *Journal of the Science of Food and Agriculture* 99(7), 3426–3433. DOI: 10.1002/jsfa.9560.

Sheikhi, A., Mirdehghan, S.H., Karimi, H.R. and Ferguson, L. (2019b) Effects of passive- and active-modified atmosphere packaging on physio-chemical and quality attributes of fresh in-hull pistachios (*Pistacia vera* L. cv. Badami). *Foods* 8(11), 564. DOI: 10.3390/foods8110564.

Slaughter, D.C., Pearson, T.C. and Studer, H.E. (1996) Hull adhesion characteristics of early-split and normal pistachio nuts. *Applied Engineering in Agriculture* 12(2), 219–221. DOI: 10.13031/2013.25642.

Steiner, W.E., Brunschweiler, K., Leimbacher, E. and Schneider, R. (1992) Aflatoxins and fluorescence in Brazil nuts and pistachio nuts. *Journal of Agricultural and Food Chemistry* 40(12), 2453–2457. DOI: 10.1021/jf00024a022.

Tavakoli, J., Sedaghat, N. and Mousavi Khaneghah, A. (2019) Effects of packaging and storage conditions on Iranian wild pistachio kernels and assessment of oxidative stability of edible extracted oil. *Journal of Food Processing and Preservation* 43(4), e13911. DOI: 10.1111/jfpp.13911.

Tavakolipour, H. (2015) Postharvest operations of pistachio nuts. *Journal of Food Science and Technology* 52(2), 1124–1130. DOI: 10.1007/s13197-013-1096-6.

Thompson, J.F. and Kader, A.A. (2016) Harvesting, transporting, processing and grading. In: Ferguson, L. and Haviland, D.R. (eds) *Pistachio Production Manual.* UCANR Publications, Richmond, California, pp. 189–194.

Thompson, J.F., Rumsey, T.R. and Spinoglio, M. (1997) Maintaining quality of bulk-handled, unhulled pistachio nuts. *Applied Engineering in Agriculture* 13(1), 65–70. DOI: 10.13031/2013.21577.

Zettler, J.L., Halliday, W.R. and Arthur, F.H. (1989) Phosphine resistance in insects infesting stored peanuts in the southeastern United States. *Journal of Economic Entomology* 82(6), 1508–1511. DOI: 10.1093/jee/82.6.1508.

Pomegranate

6

Daniel Valero Garrido[1]* and Carlos H. Crisosto[2]
[1]*Universidad Miguel Hernández, Orihuela, Spain*
[2]*University of California, Davis, California, USA*

Scientific Name, Origin and Current Areas of Production

The commercial edible pomegranate, *Punica granatum* L., belongs to the family *Punicaceae*. One of the oldest known edible fruits, it is native to Central Asia (Hummer *et al.*, 2012). The name "pomegranate" derives from the Latin name of the fruit, *mahon granatum*, which means grainy apple; it is also sometimes called Chinese apple (Patil and Karale, 1990). However, the center of pomegranate origin is Iran and Afghanistan. From there, the pomegranate spread east to India and China and west to Mediterranean countries such as Turkey, Egypt, Tunisia, Morocco, and Spain (Pareek *et al.*, 2015).

There are several types of edible pomegranate and sterile ornamental types with double flowers. Some wild-type pomegranates yield an acidic fruit, but cultivated varieties produce fruit with a sweet-sour or sweet taste. The fruit from commercial pomegranate cultivars has a leathery, smooth skin and is divided by thin, inedible membranes into several cells, each packed full of angular seeds contained in juicy pulp sacs called arils (Fig. 6.1). Fruits may be white/green, pink or red with a yellow background on the exterior; the pulp color follows that of the exterior. The arils from commercial cultivars contain a juicy edible layer that develops entirely from outer epidermal cells that elongate in a radial direction and are soft to eat (Valero and Serrano, 2010). The pomegranate fruit is nearly round, with a prominent attached calyx and a hard, leathery skin (Fig. 6.2). Surface color varies among commercial cultivars from yellow with a crimson cheek to solid brownish-red and bright red. The edible portion is the bright-red pulp (aril) surrounding the individual seed.

Spanish missionaries brought the pomegranate to the Americas in the 1500s. The optimal growing conditions for pomegranate exist in the Mediterranean region, where the fruit grows below 1,000 m altitude. This climate has a long, hot summer, mild winters, and no rain during

*Corresponding author: daniel.valero@umh.es

Fig. 6.1. Pomegranate arils. Photo courtesy of Dr. Carlos H. Crisosto.

Fig. 6.2. 'Wonderful' pomegranate with calyx. Photo courtesy of Dr. Carlos H. Crisosto.

the last stages of the fruit development, allowing fruit to mature properly. Pomegranate trees can withstand low winter temperatures and are drought and salt tolerant. However, some pomegranate cultivars grow well in tropical, subtropical, arid, and semi-arid climates. Pomegranate can withstand frost, but will not survive long at temperatures below −12°C. Areas with high relative humidity (RH) or rain in summer are totally unsuitable for its cultivation, as fruits produced under such conditions are less sweet and prone to cracking.

Currently, commercial pomegranates are cultivated extensively in Mediterranean countries (Tunisia, Turkey, Israel, Egypt, Spain, and Morocco), Iran, Afghanistan, India, and to a lesser degree in the USA (California), China, Japan, and Russia. In the USA, pomegranate production is centered in the southern San Joaquin Valley of California, predominately growing the 'Wonderful' variety. Other countries produce other cultivars such as: (i) 'Tendral' in Israel; (ii) 'Mollar' in Spain; (iii) 'Saveh sour Malas', 'Schahvar', and 'Robab' in Iran; (iv) 'Hicaznar' and 'Beynar' in Turkey; (v) 'Ganesh' in South Africa; and (vi) 'Zehri' and 'Gabsi' in Tunisia (Pareek *et al.*, 2015).

Composition and Health Benefits

The edible portion (arils) of pomegranates comprises 55–60% of the total fruit weight and consists of 80% juice and 20% seeds. The fresh juice contains 85% water with sugars, pectin, ascorbic acid, polyphenolic flavonoids, anthocyanins, and amino acids. Pomegranate juice is rich in sugars (fructose and glucose), organic acids, amino acids (glutamic and aspartic acids), vitamins, polysaccharides, and essential minerals (Pareek *et al.*, 2015). The dominant minerals in both juice and seeds are potassium, calcium, and sodium, followed by magnesium, phosphorous, zinc, iron, and copper. The concentrations of these compounds vary according to cultivar, maturity, environment, and cultivation conditions. Pomegranate fruit is a rich source of two types of polyphenolic compounds: (i) anthocyanins such as delphinidin, cyanidin, and pelargonidin, which give the fruit and juice its red color; and (ii) hydrolysable tannins such as punicalin, pedunculagin, punicalagin, gallagic and ellagic acid esters of glucose, which account for 92% of the antioxidant activity of the whole fruit. Cultivars like 'Mollar', which is commonly grown in Alicante (Spain) for eating, have soft seeds, good flavor, and arils with less pigment and acid than the juice cultivar 'Wonderful' (Fig. 6.2), which produces much more acidic, hard arils with an intense red color (Valero and Serrano, 2010). The concentration of soluble polyphenols in pomegranate juice varies between 0.2% and 1.0%. The antioxidant activity of pomegranate juice is greater than that of red wine and green tea. Pomegranate juice is an important source of phenolic compounds like anthocyanins, especially the 3-glucosides and 3,5-diglucosides of delphinidin, cyanidin, and pelargonidin. These components, along with gallagyl-type tannins, ellagic acid derivatives, and other hydrolysable tannins contribute to the antioxidant activity of pomegranate juice (Landete, 2011).

Recently, pomegranate consumption has expanded greatly due to health-promoting marketing campaigns (Valero and Serrano, 2010). Among the phytochemicals present in pomegranate are: (i) sterols and terpenoids in seeds, bark, and leaves; (ii) alkaloids in bark and leaves; (iii) fatty acids and triglycerides in seed oil; (iv) simple galloyl derivatives in leaves; (v) organic acids in juice; (vi) flavonols in the fruit rind (hull), fruit, bark, and leaves; and (vii) anthocyanins, anthocyanidins, catechin, and procyanidins in rind and juice (Patil and Karale, 1990). Historically, pomegranate bark, flowers, roots, and leaves have been used as folk remedies to treat a wide variety of diseases and ailments (Pareek *et al.*, 2015). In folk medicine, pomegranate preparations, especially of dried hulls (pericarp), roots, bark, and fruit juice were used to treat colic, colitis, and dysentery.

The potent antioxidant and antiatherogenic properties of pomegranates are attributed to their high concentration of polyphenols, including ellagic acid and punicalagin. Human studies demonstrated human absorption of ellagic acid from pomegranate juice consumption (Seeram *et al.*, 2004). Follow-up metabolic studies provided evidence that ellagic acid and ellagitannins could protect against several chronic diseases, as these ellagitannins are hydrolysed *in vivo* to ellagic acid and then transformed by the gut microbiota to urolithin derivatives, which can reduce colon cancer development by inhibiting cell proliferation and inducing cell apoptosis (Seeram *et al.*, 2004; Landete, 2011). Consumption of pomegranate juice decreases retention of harmful low-density lipoprotein (LDL). In Israel, a 3-year human study validated the hypothesis that pomegranate juice consumption by patients reduced hardening and thickening of the arteries, blood pressure, and LDL oxidation, all of which all are precursors of heart disease (Landete, 2011).

Quality and Consumer Preferences

A high-quality pomegranate should have an attractive skin color and smoothness, small seeds in the aril and should be free from sunburn (Fig. 6.3), growth cracks (Fig. 6.4), cuts, bruises, and decay (Crisosto *et al.*, 1996). Most sour and sour-sweet pomegranates have a red skin, in contrast to sweet pomegranates which have a yellow-green skin (Valero and Serrano, 2010). The acidic taste and related flavor are important attributes of pomegranate juice that contribute to its strong appeal in the food and beverage industry (Patil and Karale, 1990; Pareek *et al.*, 2015).

Fruit flavor and overall quality depend largely on the sugar and acid content of the juice. Vitamin C ranges from 0.49–30 mg 100 g^{-1} juice, depending on cultivar, which is in the low range compared with other fruit.

Fig. 6.3. Sunburn damage. Photo courtesy of Dr. Daniel Valero.

Fig. 6.4. Scars and cracks damage. Photo courtesy of Dr. Carlos H. Crisosto.

The juice content of pomegranates is 45–65% of the whole fruit or 76–85% of the aril (Valero and Serrano, 2010). The titratable acidity (TA) varies by location and year, but generally remains stable after the total soluble solids (TSS) reaches 15%. After harvest, there is no further change in either TSS or TA at 20°C. In general, cultivars with white or pink arils (such as 'Mollar' grown in Spain) are perceived as sweeter than those with purple or dark crimson arils, which tend to contain more organic acids (Pareek *et al.*, 2015). In general, the TA of current pomegranate cultivars varies between 0.13% and 5.0% at harvest.

Pomegranates are classified into three types by acidity: (i) <1% TA are sweet cultivars; (ii) 1–2% TA are sweet-sour cultivars; and (iii) >2% TA are sour cultivars. The pH varies during fruit development and ripening between 3.6 and 4.2, while minimum TSS varies from 15% to 17% (Valero and Serrano, 2010).

Fruit Physiological Characteristics

Pomegranate fruit has a very low physiological activity, measured as carbon dioxide (CO_2) and ethylene (C_2H_4) evolution that declines with time in storage. Based on the pattern of CO_2 and C_2H_4 production, pomegranate is classified as a non-climacteric fruit that exhibits no dramatic postharvest changes in physiology or composition.

Respiration activity

The ranges of fruit respiration measured as CO_2 production rates for California-grown 'Wonderful' pomegranates were 6, 12, and 24 ml CO_2 kg^{-1} h^{-1} at 5°C, 10°C and 20°C, respectively (Crisosto *et al.*, 1996). Pomegranate respiration Q_{10} values were low, ranging from 3.4 between 0°C and 10°C, 3.0 between 10°C and 20°C, and 2.3 between 20°C and 30°C (Elyatem and Kader, 1984). In the 'Gok Bache' cultivar grown in Turkey, respiration rates declined from 7.8, 4.3 and 2.4 ml CO_2 kg^{-1} h^{-1} in the first month to 0.9, 1.3 and 0.9 ml CO_2 kg^{-1} h^{-1} after 4 months of storage at 1°C, 5°C and 10°C, respectively. For South African-grown 'Herskawitz' and 'Acco', a decline in respiration rate of 67% and 68%, respectively, was observed for whole fruit when the temperature was reduced to 5°C, with an average production of 14.67 ml CO_2 kg^{-1} h^{-1}. These studies agreed with those on 'Bhagwa' and 'Ruby' cultivars, which also had lower respiration rates at harvest than during storage at 5°C and 10°C (Pareek *et al.*, 2015).

Ethylene production and sensitivity

Pomegranates produce very low amounts of ethylene: <0.1 µl kg^{-1} h^{-1} at up to 10°C and <0.2 µl kg^{-1} h^{-1} from 10°C to 20°C. Fruit are not particularly sensitive to ethylene exposure: cold or warm postharvest ethylene treatment of 'Wonderful' pomegranates caused a rapid and transient rise in CO_2, but no changes in skin color, juice color, TSS, pH, or TA (Crisosto *et al.*, 1996).

For example, exposing pomegranates to 1 µl l⁻¹ ethylene in air for 48 h briefly increased respiration and ethylene production rates, which then declined to nearly that of control fruits showing typical non-climacteric fruit behavior. These results corroborate that pomegranates do not ripen once removed from the tree and should be picked when fully ripe to ensure the best eating quality for the consumer. Postharvest ripening with or without ethylene does not add value (Valero and Serrano, 2010).

Chilling sensitivity

Most commercial pomegranates are susceptible to chilling injury (CI) and should not be stored at low temperatures (Patil and Karale, 1990; Crisosto *et al.*, 1996; Valero and Serrano, 2010; Pareek *et al.*, 2015). Specific detailed information is discussed below in the "Physiological Disorders" section.

Maturity and Harvest Indices

Fruit growth and development studies determined that the primary changes in fruit size occurred within 60 days after full bloom; thus, pomegranates are harvested when they reach a certain size and skin color (Patil and Karale, 1990; Fawole and Opara, 2013). Pomegranates do not ripen off the tree and should be picked when fully ripe to ensure the best flavor (Patil and Karale, 1990; Crisosto *et al.*, 1996; Pareek *et al.*, 2015) and postharvest life (Kashash *et al.*, 2016). The pomegranate fruit reaches full maturity (ripeness) within 4.5–6 months after full bloom, depending on cultivar and conditions. Also, there is no consistent correlation between hull skin color and aril color. Thus, hull skin color does not indicate ripening stage, harvest date, or consumption because pomegranates can attain the final color long before the arils are fully ripe. Maturity indices are cultivar-dependent and include external hull skin color (changes from yellow to red), juice color, acidity, TSS and TSS:TA ratio. The minimum maturity indices for California-grown 'Wonderful' pomegranates are red juice color equal to or darker than Munsell color chart 5R 5/12 and TA <1.9% (Crisosto *et al.*, 1996). Each cultivar requires a certain TSS:TA at harvest. TA of pomegranates varies between 0.13% and 5.0% at harvest. The TA is <1% in sweet cultivars, 1–2% in sweet-sour cultivars, and >2% in sour cultivars. TSS of pomegranates varies from 8.3% to 20.5% at harvest. Thus, maturity indices depend on cultivar. For example, TA <1.85% and TSS ≥17% are recommended for California-grown 'Wonderful' fruit, a juice tannin content <0.25% is preferred and, as already stated, red juice color equal to or darker than Munsell color chart 5R 5/12 is desirable. In Europe, where different cultivars are produced, the maturity index (MI = TSS:TA) is considered the most reliable indicator of pomegranate fruit maturity (Fawole and Opara, 2013), although its value depends on the cultivar and climate. For

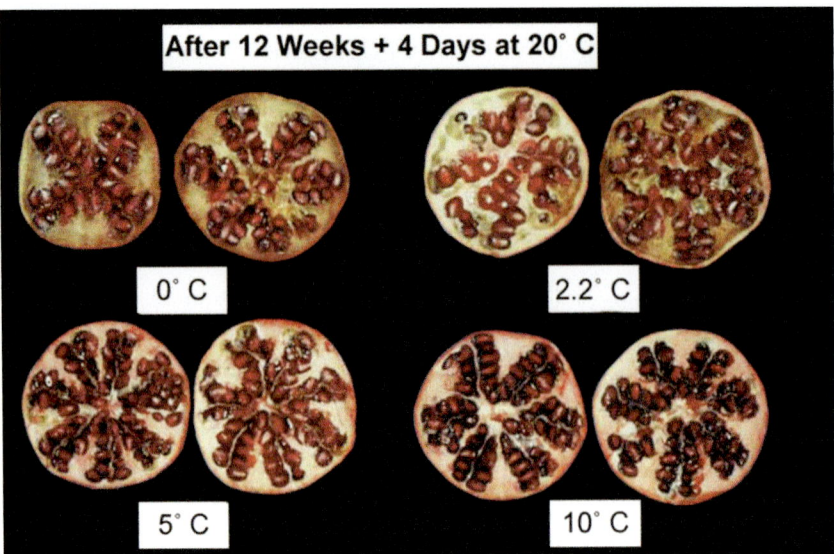

Fig. 6.5. Relationship between internal chilling injury (CI) symptoms and storage temperature. Photo courtesy of Dr. Adel A. Kader, UC Davis, California, USA.

Spanish cultivars, sweet cultivars should have an MI of 31–98; sour-sweet cultivars, 17–24; and sour cultivars, 5–7 (Valero and Serrano, 2010; Pareek *et al.*, 2015). As pomegranate cracks-splits limit late harvest, it is important to monitor harvest, orchard factors, and environmental conditions (rain or high humidity) to avoid overripe fruit and cracks-splits (Fig. 6.4).

Physiological Disorders

Chilling injury (CI)

Pomegranate fruits are very susceptible to CI if stored at temperatures between their freezing point (–3°C) and 5°C (Fig. 6.5). Thus, the minimum safe temperature for postharvest handling of pomegranates ranges between 5°C and 8°C (Crisosto *et al.*, 1996; Pareek *et al.*, 2015). For a given cultivar, the incidence and severity of CI depend on temperature and duration (Valero and Serrano, 2010). Symptoms are especially apparent upon removal of fruit from cold storage to 20°C. External CI symptoms include surface pitting, skin discoloration, scald, dead skin tissues, and increased susceptibility to decay. Internal symptoms include dead tissues, brown discoloration of the white segments separating the arils, and pale aril color (Fig. 6.5). CI also affects firmness and fruit composition, such as vitamin C and acidity concentrations (Valero and Serrano, 2010).

Fig. 6.6. Scald damage. Photo courtesy of Dr. Carlos H. Crisosto.

Husk scald

Husk scald or brown superficial discoloration is another important factor limiting the commercial storage potential of pomegranates (Fig. 6.6). Its symptoms appear as superficial skin browning initiating from the stem end of the fruit, without affecting the internal tissues, and spreading towards the blossom end as the severity increases (Parcek *et al.*, 2015). Scald incidence and severity were greater on pomegranates harvested late in the season than on those harvested mid-season, indicating that this disorder may be associated with physiological senescence. At advanced scald incidence stages, scalded areas became moldy. Husk scald may be another manifestation of CI that is a restricted to the husk. In most cultivars, scald symptoms become evident after 8 weeks of storage at 2°C or after 4–6 months at 7°C. The similarities in symptomatology and occurrence between pomegranate scald and apple scald suggest that the two disorders may have similar biochemical causes and mechanisms of control. Neither diphenylamine (DPA) nor 1-methylcyclopropene (1-MCP) treatments reduced incidence or severity of scald on 'Wonderful' pomegranates. However, controlled atmosphere (CA) storage significantly reduced scald incidence and severity on 'Wonderful' pomegranates at two maturities for up to 6 months at 7°C (Defilippi *et al.*, 2006).

In the most important Spanish cultivar, 'Mollar', scald was also reduced with 5 kPa O_2 and 15 kPa CO_2 at 5°C. This CA combination also resulted in a lower accumulation of "off flavors" than other CA treatments and decreased or prevented changes in carotenoid, acyl lipid, and phenylpropanoid metabolism that were associated with scald development in stem-end husk tissue of air-stored fruit (Valero and Serrano, 2010). Losses due to browning storage problems are greater on fruit stored at 5°C than at 2°C, possibly because of increased polyphenol oxidase (PPO) activity (Defilippi *et al.*, 2006).

Husk cracking

Husk splitting and cracking occur in fruit on the tree (Fig. 6.4). The rind shows various degrees of cracking, which often serves as entry points for decay organisms. Splitting and cracking can be prevented by using controlled, regular irrigation and avoiding late harvest or rain exposure. Rainfall on mature pomegranates following the end of the dry season can induce rapid fruit cracking similar to that of irregular irrigation. Cracking susceptibility varies among cultivars and can be reduced by regular irrigation; however, most known cultivars eventually crack if they are not harvested at commercial maturity and are overripe. Spraying with gibberellic acid (GA_3) at 150 mg l[-1] or with benzyl adenine (BA) at 40 mg l[-1] can significantly reduce cracking; however, these chemicals are not approved for commercial use. Boron application may also reduce fruit cracking (Pareek *et al.*, 2015).

Weight loss and shriveling

Pomegranates are very susceptible to water loss, resulting in shriveling of the rind. Two of the main limiting factors to extend storage of pomegranates, weight loss and shriveling, are visible only when weight loss exceeds 5% of the harvest weight. Weight loss of 'Wonderful' pomegranates during cold storage is largely due to water lost through natural porosity of the skin. Higher temperature and lower RH increase water loss because of the water pressure deficit. In general, pomegranates that are marketed without waxing should be stored at ≥90% RH to prevent water loss. However, plastic liners and waxing can reduce water loss, especially under conditions of lower RH. Modified atmosphere packaging (MAP) in plastic bags or shrink film wrapping is beneficial in reducing water loss and shrinkage and can facilitate maintenance of fruit quality for 3 months or more after harvest (Belay *et al.*, 2018).

Sunburn

This husk damage is another physiological disorder that can negatively affect commercialization of pomegranate (Fig. 6.3). The cause of sunburn is the combined action of high solar radiation, low humidity, and high temperature. Fruit surface temperatures between 41°C and 47.5°C can trigger sunburn. Kaolin clay sprays are the best method to reduce sunburn injury (Melgarejo *et al.*, 2004; Sharma *et al.*, 2018). Reduction of sunburn was reported on 'Mollar de Elche' when kaolin was sprayed over the whole canopy during maturation (Melgarejo *et al.*, 2004).

Fig. 6.7. Dumping harvested fruits from the picking bag into the bin. Photo courtesy of Dr. Carlos H. Crisosto.

Harvesting and Postharvest Handling

During hand harvesting, pickers harvest pomegranates with clippers, as thick pedicels attach the fruit strongly to branches. In some cultivars, fruit is detached by twisting the fruit pedicels off the tree. Pomegranates are selected by color and size; sometimes orchards require two harvests. Harvested fruits are placed in picking bags for transfer to harvest bins that are transported to the packinghouse (Fig. 6.7). Depending on the operation size, pomegranates can be processed and packed on arrival, bin-stored for a very short period, or cleaned, sanitized, and stored for later packing.

At the packinghouse, pomegranates are sorted to eliminate fruit with physical defects such as scuffing, cuts, bruises, splitting, and decay (Fig. 6.8). Pomegranates are highly susceptible to abrasion, impact, and compression damage, like apples. Therefore, gentle dumping, handling and transfer in the packing lines, to minimize physical damage, is essential to keep quality (Hussein *et al.*, 2019). Pomegranates are washed, air-dried to remove surface moisture, treated with fungicide, waxed, and graded (divided into several grade categories) (Fig. 6.9). Pomegranate sizing and grading can also be done optically (Fig. 6.10) prior to being packed into shipping containers. Pomegranates are usually packed into plastic (Fig. 6.11) or cardboard boxes (Fig. 6.12); and perforated plastic box liners may be used to reduce water loss during postharvest handling. Various ways to immobilize the fruits within the shipping containers may be used to reduce the incidence

Fig. 6.8. Pomegranates are sorted by hand to eliminate physical defects. Photo courtesy of Dr. Carlos H. Crisosto.

Fig. 6.9. General view of how pomegranates are graded according to size. Photo courtesy of Dr. Carlos H. Crisosto.

Fig. 6.10. Optical grading and sizing. Photo courtesy of Dr. Carlos H. Crisosto.

Fig. 6.11. Pomegranates are bulk packed into plastic boxes. Photo courtesy of Dr. Carlos H. Crisosto.

Fig. 6.12. Pomegranates packed into cardboard shipping containers. Photo courtesy of Dr. Carlos H. Crisosto.

and severity of scuffing and impact bruising during handling. In most places, pomegranates are classified into six sizes (fruit size being measured as the diameter at the widest part, which varies from 8.9 cm to 12.0 cm) and boxes are labelled according to the number of fruit they contain, ranging from 16 to 42. Thus, boxes with small-sized fruits contain a large number of fruits (e.g. 42) and boxes with large-sized fruits contain fewer fruits (e.g. 16). Only fruit with slight or no external defects are marketed fresh while fruit that have been eliminated due to physical defects (Fig. 6.8) may be used for processing into juice if the external defects are only moderate.

After processing and packing (with fungicide), pomegranates can be stored in air for up to 2 months or under CA storage for ~3 months.

Arils

Pomegranate arils are very delicate and mechanical damage is caused easily during processing by using inappropriate methods for extracting (air pressure and nozzle diameter), washing, drying to remove surface moisture, packaging, or transporting arils. This can lead to tissue wounds, abrasion, breakage, and squashing of the arils that reduces the commercial value and increases the susceptibility to decay and microbial growth. Carbon dioxide-enriched atmospheres have a fungistatic effect and their optimal range for decay control without inducing off-flavors in the arils is 15–20 kPa added to either air or 5 kPa oxygen. Although intact pomegranate

Fig. 6.13. *Botrytis* decay. Photo courtesy of Dr. Carlos H. Crisosto.

fruits are chill-sensitive, the arils are chill-tolerant and should be kept at temperatures between 0°C and 5°C to maintain quality and microbial safety (Porat *et al.*, 2016). Pomegranate arils that are not damaged or microbially contaminated can be kept at 0°C for up to 21 days, at 2°C for up to 18 days, or at 5°C for up to 14 days in marketable condition. These data are for arils extracted from freshly harvested pomegranates; a longer storage duration of intact pomegranates before aril extraction will decrease the post-extraction life of the arils.

Fruit Pathological Problems

Gray mold

Gray mold caused by *Botrytis cinerea* Pers. (Fig. 6.13) is the most economically important postharvest disease of pomegranates (Adaskaveg and Forster, 2003; Munhuweyi *et al.*, 2016). Gray mold usually starts at the calyx and as it progresses, the skin becomes light brown, tough, and leathery. Carbon dioxide-enriched atmospheres are fungistatic and inhibit growth of gray mold. Minimizing physical damage during harvesting and postharvest handling, combined with maintaining optimal temperature and RH throughout postharvest handling are very important decay control strategies for pomegranates (Nerya and Levin, 2015).

A postharvest fludioxonil immersion treatment considerably reduced postharvest decay caused by gray mold and is at present a key factor in

Fig. 6.14. Black heart decay. Photo courtesy of Dr. Themis Michailides, UC Davis-UC KARE, California, USA.

extending the postharvest life of pomegranates (Adaskaveg and Forster, 2003; Palou *et al.*, 2007). Use of fludioxonil (Scholar) as a postharvest fungicide is approved by the US Environmental Protection Agency with a maximum residue limit of 5 ppm (Adaskaveg and Forster, 2003). Natural incidence of pomegranate decay was reduced significantly to 0–8% after 2 or 5 months of storage at 10°C in fruit treated with Scholar. Scholar, a synthetic analogue of pyrrolnitrin belonging to the class of phenylpyrroles, was registered recently for controlling postharvest decay of pomegranates. An alternative approach to fungicides is using bioactive compounds distributed on the fruit (Meighani *et al.*, 2015) or arils (Martínez-Rubio *et al.*, 2015) to control decay development and improve cosmetic appearance, such as traditional edible coatings or *Aloe vera*-based coatings (Öz and Eker, 2017), and this approach is under active investigation.

Black heart and other types of decay

Pomegranate production for both fresh market and juice has increased in the last several years and black heart (Fig. 6.14) has become a disease of major concern to growers (Ezra *et al.*, 2015; Munhuweyi *et al.*, 2016; Michailides, 2020; Palou *et al.*, 2020). This disease is caused mainly by *Alternaria alternata* and other *Alternaria* spp. that develop while fruit are on the tree. *A. alternata* and related species commonly occur on plant surfaces and in dying or dead plant tissues. The pathogens overwinter on plant debris in or on the soil and in mummified fruit. The spores are airborne and can be carried to the flowers with soil dust. Infections may also start from insect and bird punctures on fruit. The fungi can grow within the fruit without external symptoms except for slightly abnormal skin color (Michailides, 2020). If the mass of blackened arils reaches the rind, it will cause softening of the affected area; these pomegranates can be detected and removed by the sorters in the packinghouse. However, the absence of clear external symptoms makes diagnosis of the disease very difficult and consumers encountering the disease may change their perception of the pomegranate's many health benefits. *Alternaria* spp. spore infections occur mainly at bloom time. Although some spore infection can occur after bloom in wounded fruit from punctures from thorns, *Hemiptera*, or cracking.

Other organisms such as *Aspergillus niger*, yeasts, *Pilidiella granati*, *Cladosporium* spp., *Penicillium crustosum* or *Coniella granati* (Kinay, 2015; Nerya *et al.*, 2016; Uysal *et al.*, 2018) can also cause pomegranate decay. Their symptoms differ only a little from those of *Alternaria* infection and their incidence is low. For example, the decay caused by *A. niger* is softer than that caused by *Alternaria* and results in exuded juice. Another common *Aspergillus* decays both arils and rind and symptoms frequently reach the outer surface of fruit, which simplifies diagnosis of the disease.

One potential control approach is fungicide coverage during bloom, but resistance has developed (Karaoglanidis *et al.*, 2011). Good orchard management practices, such as dust control and sanitation (removal of old fruit and dead branches), may reduce disease incidence (Kinay, 2015; Thomidis and Filotheou, 2016). Infected, healthy-appearing fruit may be dropped to the ground by gently shaking the tree at the time of harvest. Avoid water stress and overwatering that may result in fruit cracking. Sorting and grading of pomegranates for discoloration and cracking can help avoid packing diseased fruit.

Temperature Handling and Storage

Temperature is the environmental factor that most influences deterioration rate in pomegranate, because of its effects on fruit metabolic activity and microbial growth. For most cultivars, pomegranates should be cooled promptly after harvest. Rapid cooling down to 7°C flesh temperature is attained using forced-air cooling systems. This is one of the simplest technologies to extend postharvest life of pomegranates. After fast cooling, pomegranates should be kept at the recommended temperature during transportation and storage to reach their maximum postharvest life of 2–4 months. In general, fruit temperature must be around 7°C to prevent development of physiological disorders during storage; unfortunately, this temperature, does not prevent fungal growth. During postharvest handling, control of RH is critical because the husk desiccates readily at low RH, resulting in hard, darkened rinds that are unattractive and reduce marketability. Therefore, RH of 90–95% is preferred for storage and transportation (Crisosto *et al.*, 1996; Pareek *et al.*, 2015). Fruit waxing and storage in plastic liners can reduce weight loss. The specific optimum storage temperature varies by cultivar, production area, and postharvest handling. The recommended conditions for storage of 'Hicaznar' are 6°C with 90% RH. The cultivar 'Wonderful' grown in California and Spanish sweet pomegranates can be stored at 5°C for up to 2 months, but longer storage should be at 7°C to avoid CI (Valero and Serrano, 2010). However, at this temperature, fruit disease development becomes a problem (Adaskaveg and Forster, 2003).

Suitability as a Fresh-cut Product

Minimally processed "ready-to-eat" pomegranate arils have become popular due to their convenience, high value, unique sensory characteristics, health benefits, and ease of eating. After years of development, pomegranate arils can now be extracted by fully automated systems with minimal seed damage. The increased output and labor-cost savings are creating a new and innovative market for fresh arils, frozen arils, freeze-dried arils, dried arils, juices, wines, and health and pharmaceutical products. In addition, improved MAP technologies have been developed and tested (Belay *et al.*, 2018). Because arils have relatively low rates of respiration and ethylene production at 5°C (~ 5 ml CO_2 kg^{-1} h^{-1} and ~10 µl ethylene kg^{-1} h^{-1}), it is possible to market arils that retain good sensory and microbial quality for up to 14 days at 5°C from pomegranate fruits that were stored at 7°C for up to 3 months in air or up to 5 months in a CA of 5% O_2 + 15% CO_2 + 85% N_2 (Pareek *et al.*, 2015).

Special Storage Treatments

Controlled atmospheres (CA)

MAP and CA storage have been tested for their ability to maintain pomegranate quality during storage with satisfactory results (Mphahlele *et al.*, 2016; Porat *et al.*, 2016; Moradinezhad *et al.*, 2018). Pomegranates benefit from CA or MAP if combined with another effective treatment for decay control (Nerya and Levin, 2015; Lufu *et al.*, 2018). The combination of CA storage (5 kPa O_2 + 15 kPa CO_2) with antifungal treatment (potassium sorbate) reduced *Botrytis* decay during storage (Palou *et al.*, 2007, 2016). 'Mollar de Elche' sweet pomegranates stored at 2°C or 5°C for 12 weeks in unperforated, 25 µm-thick polypropylene film had less incidence of decay due to *Penicillium* mold (Valero and Serrano, 2010). However, more decay was found in pomegranates stored at 5°C than in those at 2°C. In

'Wonderful' pomegranate, 5 kPa O_2 + 15 kPa CO_2 and 90% RH maintained the external and internal quality for up to 6 months at 7°C without affecting flavor (Pareek *et al.*, 2015).

Heat treatment

Pomegranate arils held at 45°C for 4 min have increased antioxidant activity, which correlates with high concentrations of total phenolics and, to a lesser extent, of ascorbic acid and anthocyanins (Opara *et al.*, 2015). Additionally, the concentrations of sugars (glucose and fructose) and organic acids (malic, citric, and oxalic acids) in arils also remained at higher concentrations. Hot water dips alleviated CI by lowering electrolyte leakage and husk browning, which was attributed to the increased ratio of unsaturated:saturated fatty acids in cell membranes due to heat treatment. The degree of unsaturation of membrane lipids has been described as a mechanism of acclimation to low temperatures (Mirdehghan *et al.*, 2007).

Intermittent warming

Intermittent warming to 20°C applied in cycles, 1 day of every 6 days in storage, followed by 7 days at 15°C and 70% RH, prevent CI symptoms while maintaining pomegranate quality by reducing electrolyte and potassium ion ($K^·$) leakage, retaining anthocyanins and TA, and reducing decay in 'Mollar de Elche' pomegranates stored at 0°C or 5°C and 95% RH for 80 days (Valero and Serrano, 2010; Opara *et al.*, 2015).

Preconditioning (curing)

Exposing pomegranate to a moderate temperature (30–40°C) and high RH (90–95%) for 1–4 days considerably reduced CI and husk scald in pomegranates that were subsequently held for 80 days at 5°C or 2°C and 90–95% RH. Some limited preliminary studies indicated that methyl jasmonate (Koushesh Saba and Zarei, 2019), arginine (Babalar *et al.*, 2018), nitric acid (Ranjbari *et al.*, 2018), polyamines (Opara *et al.*, 2015), and salicyloyl (Sayyari *et al.*, 2016) were highly effective in reducing CI at 2°C for 3 months. However, none of these chemicals are currently approved for commercial use on pomegranates (Arendse *et al.*, 2015).

Retail Outlet Display Considerations

Pomegranates should not be sprinkled with water, covered with a layer of ice, or kept at low temperature to prolong shelf life. Arils packed in clamshells with atmosphere environmental control must be kept at low temperatures because of their short shelf life.

Fruit Uses

Pomegranates are primarily grown for fresh consumption of arils or juice. In smaller amounts, pomegranates are used for syrup, jams, carbonated beverages, wine coolers, and as flavoring and coloring agents in the beverage industry. In many parts of the world, pomegranate arils are used as a garnish for salads and desserts.

Cull Uses

Pomegranate juice is increasingly used as a dye in cosmetics and other products such as shampoos and high-value carpets. Hulls, pedicels, root bark, and leaves are a good source of secondary products such as tannins, dyes, and alkaloids. Potential medical uses for pomegranate hulls are being studied.

Special Research Needs

Special research needs include:

- improving aril processing and marketing;
- improving storage life by using sustainable approaches to control decay;
- searching for nutraceuticals derived from the edible and non-edible portions;
- developing cultivars with high antioxidant concentrations and flavor without CI susceptibility; and
- searching for bioactive compounds and coatings to reduce decay and deterioration.

References

Adaskaveg, J.E. and Forster, H. (2003) Management of gray mold of pomegranates caused by *Botrytis cinerea* using two reduced-risk fungicides, fludioxonil and fenhexamid. *Phytopathology* 93(6 suppl.), S 127.

Arendse, E., Fawole, O.A. and Opara, U.L. (2015) Effects of postharvest handling and storage on physiological attributes and quality of pomegranate fruit (*Punica granatum* L.): a review. *International Journal of Postharvest Technology and Innovation* 5(1), 13–31. DOI: 10.1504/IJPTI.2015.072441.

Babalar, M., Pirzad, F., Sarcheshmeh, M.A.A., Talaei, A. and Lessani, H. (2018) Arginine treatment attenuates chilling injury of pomegranate fruit during cold storage by enhancing antioxidant system activity. *Postharvest Biology and Technology* 137, 31–37. DOI: 10.1016/j.postharvbio.2017.11.012.

Belay, Z.A., Caleb, O.J., Mahajan, P.V. and Opara, U.L. (2018) Design of active modified atmosphere and humidity packaging (MAHP) for 'Wonderful' pomegranate arils. *Food and Bioprocess Technology* 11(8), 1478–1494. DOI: 10.1007/s11947-018-2119-0.

Crisosto, C.H., Mitcham, E.J. and Kader, A.A. (1996) Pomegranate: Recommendations for Maintaining Postharvest Quality. Postharvest Center, University of California, Davis, California, USA. Available at: http://post-harvest.ucdavis.edu/Commodity_Resources/Fact_Sheets/Datastores/Fruit_English/?uid=53&ds=798 (accessed 20 January 2020).

Defilippi, B.G., Whitaker, B.D., Hess-Pierce, B.M. and Kader, A.A. (2006) Development and control of scald on 'Wonderful' pomegranates during long-term storage. *Postharvest Biology and Technology* 41(3), 234–243. DOI: 10.1016/j.postharvbio.2006.04.006.

Elyatem, S.M. and Kader, A.A. (1984) Post-harvest physiology and storage behaviour of pomegranate fruits. *Scientia Horticulturae* 24(3-4), 287–298. DOI: 10.1016/0304-4238(84)90113-4.

Ezra, D., Kirshner, B., Gat, T., Shteinberg, D. and Kosto, I. (2015) Heart rot of pomegranate. When and how does the pathogen cause the disease? *Acta Horticulturae* 1089, 167–171.

Fawole, O.A. and Opara, U.L. (2013) Developmental changes in maturity indices of pomegranate fruit: a descriptive review. *Scientia Horticulturae* 159, 152–161. DOI: 10.1016/j.scienta.2013.05.016.

Hummer, K.E., Pomper, K.W., Postman, J., Graham, C.J., Stover, E. *et al.* (2012) Emerging fruit crops. In: Badenes, M.L. and Byrne, D.H. (eds) *Fruit Breeding.* Springer, New York, NY, pp. 97–147.

Hussein, Z., Fawole, O.A. and Opara, U.L. (2019) Bruise damage susceptibility of pomegranates (*Punica granatum*, L.) and impact on fruit physiological response during short term storage. *Scientia Horticulturae* 246, 664–674. DOI: 10.1016/j.scienta.2018.11.026.

Karaoglanidis, G.S., Luo, Y. and Michailides, T.J. (2011) Competitive ability and fitness of *Alternaria alternata* isolates resistant to QoI fungicides. *Plant Disease* 95(2), 178–182. DOI: 10.1094/PDIS-07-10-0510.

Kashash, Y., Mayuoni-Kirshenbaum, L., Goldenberg, L., Choi, H.J. and Porat, R. (2016) Effects of harvest date and low-temperature conditioning on chilling tolerance of 'Wonderful' pomegranate fruit. *Scientia Horticulturae* 209, 286–292. DOI: 10.1016/j.scienta.2016.06.038.

Kinay, P. (2015) Alternative technologies to control postharvest diseases of pomegranate. *Stewart Postharvest Review* 11: Article 3.

Koushesh Saba, M. and Zarei, L. (2019) Preharvest methyl jasmonate's impact on postharvest chilling sensitivity, antioxidant activity, and pomegranate fruit quality. *Journal of Food Biochemistry* 43(3): e12763. DOI: 10.1111/jfbc.12763.

Landete, J.M. (2011) Ellagitannins, ellagic acid and their derived metabolites: a review about source, metabolism, functions and health. *Food Research International* 44(5), 1150–1160. DOI: 10.1016/j.foodres.2011.04.027.

Lufu, R., Berry, T.M., Ambaw, A. and Opara, U.L. (2018) The influence of liner packaging on weight loss and decay of pomegranate fruit. *Acta Horticulturae* 1201, 259–264. DOI: 10.17660/ActaHortic.2018.1201.35.

Martínez-Rubio, D., Pina, N., Guillén, F., Valero, D., Zapata, P.J. *et al.* (2015) Application of an edible coating based on *Aloe vera* to improve general quality of minimal processed pomegranate arils. *Acta Horticulturae* 1071, 489–494. DOI: 10.17660/ActaHortic.2015.1071.63.

Meighani, H., Ghasemnezhad, M. and Bakhshi, D. (2015) Effect of different coatings on post-harvest quality and bioactive compounds of pomegranate (*Punica granatum* L.) fruits. *Journal of Food Science and Technology* 52(7), 4507–4514. DOI: 10.1007/s13197-014-1484-6.

Melgarejo, P., Martínez, J.J., Hernández, F., Martínez-Font, R., Barrows, P. *et al.* (2004) Kaolin treatment to reduce pomegranate sunburn. *Scientia Horticulturae* 100(1-4), 349–353. DOI: 10.1016/j.scienta.2003.09.006.

Michailides, T. (2020) Black Heart of Pomegranate. Kearney Agricultural Research and Extension Center, University of California, Parlier, California, USA. Available at: http://kare.ucanr.edu/programs/Plant_pathology/Black_Heart_of_Pomegranate/index.cfm (accessed 3 January 2020).

Mirdehghan, S.H., Rahemi, M., Martínez-Romero, D., Guillén, F., Valverde, J.M. *et al.* (2007) Reduction of pomegranate chilling injury during storage after heat treatment: role of polyamines. *Postharvest Biology and Technology* 44(1), 19–25. DOI: 10.1016/j.postharvbio.2006.11.001.

Moradinezhad, F., Khayyat, M., Ranjbari, F. and Maraki, Z. (2018) Physiological and quality responses of Shishe-Kab pomegranates to short-term high CO_2 treatment and modified atmosphere packaging. *International Journal of Fruit Science* 18(3), 287–299. DOI: 10.1080/15538362.2017.1419399.

Mphahlele, R.R., Fawole, O.A. and Opara, U.L. (2016) Influence of packaging system and long term storage on physiological attributes, biochemical quality, volatile composition and antioxidant properties of pomegranate fruit. *Scientia Horticulturae* 211, 140–151. DOI: 10.1016/j.scienta.2016.08.018.

Munhuweyi, K., Lennox, C.L., Meitz-Hopkins, J.C., Caleb, O.J. and Opara, U.L. (2016) Major diseases of pomegranate (*Punica granatum* L.) their causes and management – a review. *Scientia Horticulturae* 211, 126–139. DOI: 10.1016/j.scienta.2016.08.016.

Nerya, O. and Levin, A. (2015) Innovative treatment of pomegranates from harvest to market. *Acta Horticulturae* 1089, 489–493. DOI: 10.17660/ActaHortic.2015.1089.67.

Nerya, O., Tzviling, A., Hibrahim, H. and Ben-Arie, R. (2016) *Coniella granati* – a new pathogen of pomegranates in Israel – postharvest fungicide screening for control of storage decay. *Acta Horticulturae* 1144, 465–468. DOI: 10.17660/ActaHortic.2016.1144.69.

Opara, U.L., Atukuri, J. and Fawole, O.A. (2015) Application of physical and chemical postharvest treatments to enhance storage and shelf life of pomegranate fruit – a review. *Scientia Horticulturae* 197, 41–49. DOI: 10.1016/j.scienta.2015.10.046.

Öz, A.T. and Eker, T. (2017) Effects of edible coating of minimally processed pomegranate fruit. *Journal of Horticulture, Forestry and Biotechnology* 21, 105–109.

Palou, L., Crisosto, C.H. and Garner, D. (2007) Combination of postharvest antifungal chemical treatments and controlled atmosphere storage to control gray mold and improve storability of 'Wonderful' pomegranates. *Postharvest Biology and Technology* 43(1), 133–142. DOI: 10.1016/j.postharvbio.2006.08.013.

Palou, L., Rosales, R., Montesinos-Herrero, C. and Taberner, V. (2016) Short-term exposure to high CO_2 and O_2 atmospheres to inhibit postharvest gray mold of pomegranate fruit. *Plant Disease* 100(2), 424–430. DOI: 10.1094/PDIS-06-15-0637-RE.

Palou, L., Kinay-Teksür, P., Cao, S., Karaoglanidis, G. and Vicent, A. (2020) Pomegranate, persimmon, and loquat. In: Palou, L. and Smilanick, J.L. (eds) *Postharvest Pathology of Fresh Horticultural Produce.* CRC Press, Boca Raton, Florida, USA, pp. 187–225.

Pareek, S., Valero, D. and Serrano, M. (2015) Postharvest biology and technology of pomegranate. *Journal of the Science of Food and Agriculture* 95(12), 2360–2379. DOI: 10.1002/jsfa.7069.

Patil, A.V. and Karale, A.R. (1990) Pomegranate. In: Kose, T.K. and Mitra, S.K. (eds) *Fruits: Tropical and Subtropical.* Naya Prokash, Calcutta, India, pp. 614–631.

Porat, R., Kosto, I. and Daus, A. (2016) Bulk storage of 'Wonderful' pomegranate fruit using modified atmosphere bags. *Israel Journal of Plant Sciences* 63(1), 45–50. DOI: 10.1080/07929978.2016.1152839.

Ranjbari, F., Moradinezhad, F. and Khayyat, M. (2018) Effect of nitric oxide and film wrapping on quality maintenance and alleviation of chilling injury on pomegranate fruit. *Journal of Agricultural Science and Technology* 20, 1025–1036.

Sayyari, M., Aghdam, M.S., Salehi, F. and Ghanbari, F. (2016) Salicyloyl chitosan alleviates chilling injury and maintains antioxidant capacity of pomegranate fruits during cold storage. *Scientia Horticulturae* 211, 110–117. DOI: 10.1016/j.scienta.2016.08.015.

Seeram, N.P., Lee, R. and Heber, D. (2004) Bioavailability of ellagic acid in human plasma after consumption of ellagitannins from pomegranate (*Punica granatum* L.) juice. *Clinica Chimica Acta* 348(1-2), 63–68. DOI: 10.1016/j.cccn.2004.04.029.

Sharma, R.R., Datta, S.C. and Varghese, E. (2018) Effect of Surround WP®, a kaolin-based particle film on sunburn, fruit cracking and postharvest quality of 'Kandhari' pomegranates. *Crop Protection* 114, 18–22. DOI: 10.1016/j.cropro.2018.08.009.

Thomidis, T. and Filotheou, A. (2016) Evaluation of five essential oils as biofungicides on the control of *Pilidiella granati* rot in pomegranate. *Crop Protection* 89, 66–71. DOI: 10.1016/j.cropro.2016.07.002.

Uysal, A., Kurt, Ş., Soylu, E.M., Kara, M. and Soylu, S. (2018) Morphology, pathogenicity and management of Coniella fruit rot (*Coniella granati*) on pomegranate. *Turkish Journal of Agriculture - Food Science and Technology* 6(4), 471–478. DOI: 10.24925/turjaf.v6i4.471-478.1787.

Valero, D. and Serrano, M. (2010) *Postharvest Biology and Technology for Preserving Fruit Quality.* CRC Press, Boca Raton, Florida, USA.

Table Grape

7

Carlos H. Crisosto[1]*, Reinaldo Campos-Vargas[2], and Amnon Lichter[3]
[1]*University of California, Davis, California, USA*
[2]*Universidad Andrés Bello, Santiago, Chile*
[3]*The Volcani Center, Rishon LeZion, Israel*

Scientific Name, Origin and Current Areas of Production

Table grapes (*Vitis vinifera* L.) are deciduous, woody vines native to the Mediterranean and Central Asia. They produce fruits for fresh consumption that botanically are clusters of true berries. Berries develop as a part of a cluster inflorescence composed of a rachis as the main herbaceous axis and two prominent and three or four smaller rachis ramifications. Clusters are attached to the vine by the peduncle and each berry is attached to the cluster by a pedicel. The berry is derived from a single ovary, after the fusion of two carpels surrounded by the ovary wall that ripen into an edible pericarp. In most commercial cultivars, each carpel can contain two seeds (in seeded cultivars) or seed traces due to ovule abortion in some seedless (stenospermocarpic) cultivars. Berries are composed of exocarp (skin and cuticle), mesocarp (flesh), and endocarp (tissue around the seeds) tissues. Table grape production comprises nearly 36% of total world grape production, which also includes wine grapes and raisins. Table grapes are mainly produced in Mediterranean areas, but are also grown in tropical and subtropical regions. The major grape-producing countries are China (9.5 million t), Turkey (1.9 million t), India (1.9 million t), Iran (1.7 million t), Egypt (1.5 million t), Uzbekistan (1.2 million t), Italy (1.1 million t), the USA (1 million t), Brazil (0.8 million t), Chile (0.7 million t), Peru (0.6 million t), Mexico (0.4 million t), South Africa (0.3 million t), Greece (0.3 million t), Spain (0.3 million t), and Australia (0.1 million t) (International Organisation of Vine and Wine, 2019).

---

*Corresponding author: chcrisosto@ucdavis.edu

Maturity and Harvest Indices

Total soluble solids (TSS), color, and/or titratable acidity (TA) are used as maturity indices for table grapes because they predict consumer acceptance and potential postharvest life (Thompson and Crisosto, 2000). TSS is an indirect measure of sugar content, determined using a refractometer. TA is measured by titration with 0.1 N sodium hydroxide to an end point of pH 8.2, using the color change of phenolphthalein as an indicator. There are portable machines that can measure both parameters (TSS and TA) automatically, but which need careful calibration. These maturity indices guide harvest according to cultivar, production area, and specific marketing regulations. Most cultivars are harvested at a berry TSS of 15–17% and TA<0.6%. Thus, in early-production areas, a TSS:TA ratio of 20 or greater is used as a minimum maturity index. Many green cultivars (e.g. 'Thompson Seedless') turn yellow when ripe. Cultivars that are not green-colored, mostly red or black, have minimum color requirements based on the percentage of berries in the cluster that show a minimum color intensity and coverage.

Consumer Quality

Consumers expect fresh grapes to be free of defects such as decay, cracked berries, stem browning, shriveling, sunburn, dried berries, or insect damage. High consumer acceptance is attained for fruit with high TSS content or a high TSS:TA ratio. Some cultivars have low acidity (0.2–0.3%) and are perceived as sweet at a TSS of 15%, but have a "flat" taste. Other cultivars have high TA (0.5–0.6%) and require greater TSS to gain high consumer acceptance. For example, 'Red Globe' grapes with TSS<16.1% and a TSS:TA of 20.1–22.5 were liked by Caucasian–Hispanic US consumers (76% satisfied), but fewer Chinese consumers liked them (42%). Similarly, the category of "neither like nor dislike" was chosen by only 6% of US consumers, compared with 38% of Chinese consumers, reflecting the fact that Chinese consumers prefer grapes with lower TA. Berry texture, perceived by consumers as crunchiness and measured by firmness, is an important factor for consumer acceptance, as is skin thickness (Crisosto and Crisosto, 2002). Skin color can influence consumer acceptance: for example, Asian markets prefer red, but not dark-red 'Crimson Seedless' grapes. These studies reflect important ethnic differences in consumer expectations. As new cultivars with improved flavor become available and new global markets open, there is a growing demand to tailor fruit quality to specific markets (Wang *et al.*, 2017).

Fruit Physiological Characteristics

Table grapes are classified as non-climacteric fruit. They should be harvested at full maturity, since their quality will not improve (and begins to deteriorate) after harvest. Table grapes have a relatively low respiration rate. Table grapes produce very low amounts of ethylene and carbon dioxide. Respiration measured as carbon dioxide (CO_2) evolution depends on the cultivar and temperature.

Respiration and carbon dioxide production

Respiration was measured in popular table grape cultivars and it varied from 3 ml CO_2 kg^{-1} h^{-1} at 0°C, increasing to 7 ml CO_2 kg^{-1} h^{-1} at 5°C, and to 14 ml CO_2 kg^{-1} h^{-1} at 10°C. The respiration rate of specific cluster components was measured using a calorimeter at 4°C. The respiration rate of the rachis alone is ~15 times greater than that of the berry (Thompson and Crisosto, 2000; Silva-Sanzana et al., 2016). Table grapes produce very little ethylene, with a peak at bloom followed by decreasing concentrations until harvest. In some cultivars, a very small rise in ethylene evolution has been detected at veraison, but the potential role of this small ethylene peak in maturity development is not fully understood. However, exogenous application of ethephon to red and black cultivars at or after veraison enhances color development, increases TSS, reduces acidity and firmness; and may induce berry abscission and softening. The role of abscisic acid (ABA) in grape maturity is under investigation (Pilati et al., 2017). ABA accumulates transiently after veraison and its exogenous application enhances color development and maintains postharvest quality of 'Crimson Seedless', 'Flame Seedless' and 'Red Globe' grapes, without the increased softening and reduced storage potential that are side effects of ethephon treatment (Cantín et al., 2007).

Ethylene production and sensitivity

Table grapes produce <0.1 µl kg^{-1} h^{-1} ethylene at 20°C. Grapes are not very sensitive to ethylene exposure during storage (Palou et al., 2003). However, recent work suggested that susceptibility to ethylene could be cultivar dependent (Li et al., 2015), thus, further work should be carried out to understand the role of ethylene in grape postharvest quality.

Water loss

Berries are composed of three types of tissues: (i) the exocarp; (ii) the mesocarp (flesh: ~80% of berry weight); and (iii) the endocarp (tissue around the seeds). The exocarp includes the cuticle (cutin, waxes, and soluble lipids) and skin (epidermis and hypodermis). The epidermis is composed of a few cell layers where most of the pigments (anthocyanins, carotene, and chlorophyll) and flavor and aroma compounds accumulate. The hypodermis exhibits periclinal and anticlinal growth, allowing three-dimensional enlargement of the berry. Initially, grape skin contains stomata, but as the

Fig. 7.1. Different levels of rachis browning damage due mainly to water loss during handling. Photo courtesy of Dr. Carlos H. Crisosto.

berry matures, stomate density decreases and suberized lenticels remain by the end of the fruit-growing stage. Therefore, berry water loss occurs mainly through the cuticle. The cuticle is composed of cutin, waxes, and soluble lipids and provides the primary barrier limiting water transport, respiration, and water loss; also it confers resistance to decay fungi. Its formation begins about 3 weeks before anthesis as a highly organized tissue, but the cuticle content decreases during ripening, making table grapes highly susceptible to water loss when berries are exposed to high temperature and air velocities, or low humidity conditions.

Cumulative water loss during postharvest handling results in weight loss, rachis browning (Fig. 7.1), berry shatter, and even shriveling of berries (Crisosto *et al.*, 2001; Lichter, 2016). There is a strong correlation between cluster water loss and rachis browning (Bahar *et al.*, 2017). The high respiration rate and small diameter of rachis tissue may also contribute to browning and it may be reduced by the use of 1-methylcyclopropane (1-MCP) (Li *et al.*, 2015). Clusters from some cultivars are prone to rachis browning when shoulder and central rachis thicknesses are less than 3 mm and 2.5 mm, respectively. Water loss of 0.5–2.1% can occur during an 8 h wait before cooling. The magnitude of water loss is directly related to exposure time, ambient temperature, and type of box material. Even a delay of a few hours can cause severe drying and browning of cluster rachises,

Fig. 7.2. Berry browning in 'Italia' grapes due to chilling injury. Photo courtesy of Dr. Carlos H. Crisosto.

especially on the hottest days. However, rachis browning is not observed at harvest: the damage will be displayed after a week of cold storage if water loss exceeded 2% during delayed cooling (Crisosto *et al.*, 2001). During postharvest life, the temperature and relative humidity (RH), expressed as water pressure deficit (WPD), air velocity and time of exposure, are the main physical factors that contribute to water loss. Therefore, short cooling delays and fast cooling are highly recommended. When the fruit is packed into a box and/or polyethylene bags, the humidity inside the bag increases and WPD is reduced, but fruit temperature remains a critical factor. The time it takes for the clusters to achieve the storage room temperature (delay in cooling) is critical to reduce WPD; therefore, fast hauling (field to cold storage) and forced-air cooling systems have become standard practices in the table grape industry.

Chilling sensitivity

Most table grape cultivars are not sensitive to chilling, and their postharvest life can be extended using low temperature storage near rachis freezing points. Some cultivars, especially Muscat types such as 'Princess' or 'Italia', exhibit damage such as 'berry browning' (Fig. 7.2) when exposed to low temperatures that are well above their freezing point (Vial *et al.*, 2005). In these cultivars, the flesh browning disorder that is restricted to the skin is associated with cell membrane damage induced by chilling injury.

Packaging

There are four types of box materials commonly used for table grapes (Luvisi *et al.*, 1995; Thompson and Crisosto, 2000): (i) corrugated cardboard, with or without a coating; (ii) expanded polystyrene (EPS or Styrofoam); (iii) plastic; or (iv) wood. In global markets, most shipping boxes are transported on 1.2 m × 1.0 m pallet bases (Euro-pallets). Therefore, package sizes for a large portion of the market must be compatible with metric shipping containers. To maximize pallet area use and assure airflow through the fruit, there are different loading patterns for each container type (metric: 40 cm × 30 cm; shoe box: 50 cm × 30 cm; Euro: 60 cm × 40 cm; or metric-ESP: 50 cm × 40 cm, all loaded 10 –13 cm high). The pallet height cannot be more than 1.8 m, with 72–96 boxes per pallet depending on box dimensions, to assure airflow and remain under the maximum weight for the vehicle. The vented area (VA) of these containers, which is related to sulfur dioxide pad performance, varies widely from ~3.3% to 18.2%. Box material is also related to performance, as sulfur dioxide gas absorption depends on the box material. The choice of box material is often influenced by factors other than maintaining the quality of packed fruit, such as receiver preferences, environmental issues (recycling), cost, cold-storage humidity conditions, length of storage period, box weight, etc. Plastic and EPS (Styrofoam) containers are becoming more popular for late-season cultivars and long storage periods because they maintain their structural integrity under high humidity conditions better than corrugated boxes. A large portion of the total grape production is packed in clamshell containers. In the USA, mainly due to retail pressures to use the 1.0 m × 1.2 m Euro-pallet, the range of box sizes was diversified well beyond the old standard LA lug (35.6 cm × 44.4 cm).

Handling During Harvesting and Packing

There are two main systems used commercially to pack table grapes: (i) field packing; and (ii) shed packing (Thompson and Crisosto, 2000). Field packing is the main practice in California, while packing in sheds is the main practice in South Africa and Chile. In Italy, field packing is the preferred method during the early season, but shed packing is used for mid- and late-season cultivars. A recent innovation in shed packing is for some initial cleaning and trimming to occur in the field, prior to transport to the shed for final packing.

Field-packing system

The most common field-packing system in California is the 'avenue pack' (Fig. 7.3). In this system, grape clusters are carefully picked, trimmed to

Fig. 7.3. Avenue packing, general field view. Photo courtesy of Dr. Carlos H. Crisosto.

reduce visual quality losses (carried out by trained pickers holding clusters away from the main rachis), and placed into shallow plastic picking lugs. The bloom (natural wax) on the grape berry's surface has a typical structure of ridges and is a primary appearance quality factor. Rough handling and berry rubbing destroys this bloom, giving the skin a shine rather than the more desirable luster. The picking containers are then transferred a short distance to the packer, who works at a small, shaded portable stand at the end of the avenue between vineyard blocks (Fig. 7.4). It is common for one packer and two to three pickers to work as a crew. Packing materials are located at the packing stand. With many packing stands around the vineyard, supervision is more critical than in a packing shed. Lidding is also done in the field. Substandard quality grapes (culls) can be accumulated in field containers for transport to wineries or other processors. After packing and lidding, boxes of grapes are loaded on to disposable or recycled pallets and placed in the shade of the vines to await fast transport to the shed. These boxes are loaded in a pallet unit (unitized) so they are easy to handle, and they are usually held together by strapping or netting. Modified small or large trucks using shade screens collect the pallets frequently. Some strapping is necessary in the field before loading on the truck, especially for grapes packed in shoe boxes or EPS (Styrofoam) boxes.

Upon arrival at the cold storage facility, loaded pallets coming from the field often pass through a "pallet squeeze" (Fig. 7.5), a device that straightens and tightens the stacks of containers, and under best practices

Fig. 7.4. Packer, working in a small, shaded portable stand as part of the avenue packing system, places grapes in a restricted cluster bag (RCB) before packing the grapes into EPS boxes. Photo courtesy of Mr. Mike Poe, UC ANR, Davis, California, USA.

are moved to cooling and sulfur dioxide fumigation facilities within the first 6 h after harvest. There, forced-air cooling begins immediately and reduces grape temperature to near 0°C. Grapes do not tolerate the wetting associated with hydrocooling.

Shed-packing system

This handling system is recommended when no well-trained field crews are available, as trained and experienced quality control is performed in the comfortable, controlled-environment conditions of a packing shed (Fig. 7.6). In the shed-packing system, there are two variations according to when trimming and cleaning occur. In most shed-packing operations in Chile and California, a first quality sorting that includes trimming, color sorting, and sizing occurs in the field and a second trimming, sorting, and sizing operation occurs in the packing shed; in other cases quality sorting only takes place once in the packing shed (Fig. 7.7). Trained operators carefully select mature grapes and avoid mechanical berry damage by gently using small sharp scissors with round tips during harvesting. A field trimming to remove defective or decayed berries and to obtain a better cluster shape is highly recommended, then clusters are placed carefully into plastic harvesting boxes or field lugs (60 cm × 40 cm × 25 cm). The

Fig. 7.5. Pallet squeezing to straighten and tighten the stacks of containers. Photo courtesy of Mr. Mike Poe, UC ANR, Davis, California, USA.

Fig. 7.6. Shed packing, general view. Photo courtesy of Dr. Carlos H. Crisosto.

Fig. 7.7. Grading, cleaning, and sizing during a shed-packing operation. Photo courtesy of Dr. Carlos H. Crisosto.

field lugs are then transported to small sheds or a large packinghouse with temperature control and fast cooling facilities. At the packing shed, the field lugs are distributed to packers who grade and pack the fruit. Often, two to three different grades are packed simultaneously by each packer to facilitate quality selection.

In most places, the shed packaging is typically carried out in large packinghouses, under controlled temperature conditions and adequate lighting. Plastic boxes filled with clusters are placed on to a packing belt from which workers start or continue preparing the clusters. Workers are located at a "working table" where the clusters are selected and transferred into clean lugs. Trimming is done at the working station in the packing shed, and damaged and small berries left from the cleaning and trimming operations are collected in individual containers located further down the working table or on a conveyer belt that moves the waste out of the area. After this, grapes are fumigated with sulfur dioxide (SO_2), then moved to the packing section.

In the packing section, workers are allocated to a packaging table where they take the lug containing clean, selected, trimmed, and SO_2-fumigated clusters from the conveyer and pack clusters into different shipping boxes, according to cluster shape, berry color, and cluster size. In a separate line, located over the packing table workstation, a roller conveyer moves empty shipping boxes. Packaging materials, such as cartons, wooden boxes, waxed corrugated paper boxes, ESP, non-returnable plastic boxes,

SO$_2$-generating pads and plastic or paper bags, are provided at the packing table workstation.

Details of packing

In both field-packing and shed-packing systems, a weighing scale is provided to control the total box weight and, when needed, dual-release SO$_2$-generating pads are placed on the top of the box. Grapes are nearly always packed on a scale to facilitate packing to a precise net weight; in all cases, packed lugs are subject to quality inspection and weight checking. Typically, mid- and late-season grapes are packed in plastic bags or wrapped in paper, while early-season grapes are usually in bulk packs. In all cases, packed lugs are subject to quality inspection and weight checks. Each packaged box should include the necessary information to track the grapes from vineyard to market. Therefore, the producer brand name, geographic location, product name, and table grape cultivar should be identified on each shipping box. The weight in kilograms and date of packaging must be indicated as well. Postharvest treatments such as sulfur dioxide applications must be marked clearly on the container to comply with market regulations.

Handling challenges

Handling challenges are faced when the production area is far from the packing shed–cold storage facility or where fast cooling facilities are unavailable. In this case, trimmed clusters in field lugs or packed grapes after harvest should not be cooled down to 0°C initially, but instead only down to ~8–10°C to avoid condensation on the berry surface. The presence of free water on the berry surface triggers the onset of *Botrytis* mold development. Optimum temperature management will depend on facility conditions, packaging materials, and environmental conditions that determine the dew point (condensation); thus, temperature management must be fine-tuned to protect table grapes under this situation and handling should be constantly evaluated for decay development and visual quality losses.

Post-packing Handling

Under both field-packing and shed-packing systems, after palletization is completed, the pallets are moved for immediate SO$_2$ treatment either to a warm fumigation chamber, forced-air cooler/fumigation chamber, or just to a forced-air cooler (Fig. 7.8), with fumigation at the end of the packing day. In all systems, cooling must start as soon as possible and SO$_2$ must be applied within 6–12 h of harvest.

Many grape forced-air coolers in California (Fig. 7.8) are designed to achieve seven-eighths of the cooling time in 6–10 h. After cooling is

Fig. 7.8. Forced-air cooling of table grapes packed in cardboard boxes. Photo courtesy of Mr. Mike Poe, UC ANR, Davis, California, USA.

completed, the pallets are moved to a storage room to await transport. Ideally, the air storage room operates at −1–0°C and 90–95% RH, with a moderate airflow of 0.01–0.02 m³ min⁻¹ t⁻¹ of stored grapes (J.F. Thompson, California, USA, 2019, personal communication). Constant low temperature, high RH, and moderate airflow are important to limit the rate of water loss from berries and rachises. Fruit should be stored at −0.5–0°C pulp temperature throughout its postharvest life. Lower air and/or berry temperatures can damage the rachis and trigger browning.

Optimum Storage Conditions

Ideally, initial cooling (precooling) must start within 6 h and sulfur dioxide must be applied no later than 12 h after harvest. Generally, the forced-air cooling system should be designed to deliver an airflow of 0.05 m³ min⁻¹ t⁻¹. After cooling is completed, pallets are moved to a storage room to await distribution (Fig. 7.9). A berry storage temperature of −0.5–0°C with a RH of 90–95% and an airflow of 0.01–0.02 m³ min⁻¹ t⁻¹ is recommended during long-term storage. The constant low temperature, high RH, and moderate airflow conditions will limit weight loss, rachis browning, and disease development. A relationship between air velocity and RH on fruit weight loss has been described during storage. Weight loss was not affected by air velocity (3.7–30.5 m min⁻¹) when fruit were exposed to high RH (95%).

Fig. 7.9. Typical Californian table grape storage using racks. Photo courtesy of Mr. Mike Poe, UC ANR, Davis, California, USA.

However, when fruit is exposed to RH below 85% and high air velocity it has ~1.5% more weight loss than fruit stored at high RH. Thus, high RH overcomes the negative effect of high air velocity. Freezing damage may occur in less-mature grapes when temperature is at the low end of the range. The highest freezing point for berries is between −2°C and 3°C, but the freezing point varies depending on TSS. However, a −2°C freezing point for the rachis has been reported for wine grapes, and some new table grape cultivars also may be more sensitive to rachis freezing damage. Freezing damage is visible as external browning and a milky color in the berry. Pedicels (capstems) and laterals or the main axis on the clusters are also affected by freezing temperatures with similar browning symptoms of water loss.

Transportation to markets

Shipping of table grapes requires selection of a marine container, cold storage in the vessel, or truck transportation system. Currently, 30–40 days transportation are required from origin to final destination in common routes from South Africa and South America to Europe or the USA (R. Campos-Vargas, Chile, 2019, personal communication). The environmental conditions described above are critical to maintain grape quality and reduce claims at arrival. Prior to loading for transportation, it is important to assure uniform and optimal berry temperature among boxes and pallets in the load. In Chile, pallets are stored in a cold room for 3–4 days to homogenize fruit temperature before loading into the container or vessel at the port.

During this temperature homogenization treatment, the air speed is set at 1.0–1.5 m s^{-1} compared with an air speed of 0.1–0.2 m s^{-1} during standard storage. In California, as forced-air cooling, storage, and loading of marine containers occur in the same cold storage location, homogenizing berry temperature within and across pallets is not necessary. Some operators may use forced-air cooling on loads when lengthy inspections or re-packaging and cleaning occurred before shipment.

Retail Outlet Display Considerations

Upon arrival at retail stores, grapes should be refrigerated immediately. The ideal condition for long-term display of grapes is 0°C with 90–95% RH (Thompson and Crisosto, 2000). Grapes should be kept away from water or ice, as moisture will decrease their shelf life. Grape boxes should be stacked in the cooler so that air can circulate around them, but dehydration, especially of rachises, will accelerate if they are stored near the cooling unit's direct air path. Grapes tend to absorb odor and should not be stored next to green onions, garlic, or leeks (Thompson and Crisosto, 2000). Stacked grape boxes should be kept off the ground to avoid damage from any excess moisture or a dirty surface. There is marketing information that supports increased sales with a large, unrefrigerated display area (Fig. 7.10) selling grapes of different colors. Clusters of the three main cultivars were placed at different retail display temperatures after a 7-day cold-storage period to simulate fast shipping, distribution, and delivery to the retail store. Display temperature did not affect shrinkage until after 48 h display, suggesting that display rotation should start after 48 h. After 48 h, grapes displayed at 5°C generally had less shrinkage than grapes displayed at 20°C or 25°C. Packaging also affected shelf life. The percentage shrinkage of grapes packed in slider bags and displayed at 5°C was significantly greater at 72 h compared with 24 h or 48 h, while that of grapes packed in slider bags displayed at 20°C or 25°C was significantly greater at 48 h than 24 h. Grapes packed in clamshells (0.90 kg and 1.8 kg) and displayed at 5°C did not experience much shrinkage within the 72 h retail display period, but if displayed at ≥20°C, the percentage shrinkage was significantly greater at 72 h.

Physiological Disorders

Several symptoms of grape deterioration are classified as physiological disorders although these disorders cause significant losses during grape marketing (Cappellini *et al.*, 1986; Thompson and Crisosto, 2000; Blanckenberg

Fig. 7.10. Large unrefrigerated display of table grapes in a retail store. Photo courtesy of Dr. Carlos H. Crisosto.

et al., 2018). The losses have proved difficult to study as incidences are erratic. The current state of knowledge is discussed below.

Shatter

Shatter is loss of berries from the capstems and is expressed as a percentage (loose berry weight divided by the total weight of the grapes per box × 100). In the USA, a maximum 3.0% shatter at harvest is currently accepted for high-quality California grapes. In most cultivars, shattering is determined early during fruitlet development by the formation of an abscission layer at the capstem (the insertion point of the berry into the pedicel). In a few cultivars, the abscission layer does not occur at the capstem and berries shatter with pedicles. A 'wet' shatter where berries separate, leaving part of the vascular bundle on the pedicel, is associated with rough handling of

Fig. 7.11. Table grapes affected by waterberry. Photo courtesy of Dr. Carlos H. Crisosto.

'Thompson Seedless'. Orchard factors during ripening affect shatter: for instance, shatter increases with gibberellic acid (GA_3) and/or N-(2-chloro-4-pyridyl)-N-phenylurea (CPPU) applications for berry enlargement, due to loss of rachis flexibility (Avenant, 2017). Shatter is aggravated by rough handling during packing (Luvisi *et al.*, 1995), with additional damage occurring through final retail sale, and increases in severity the longer fruit remain on the vine (Thompson and Crisosto, 2000). Shatter susceptibility varies among cultivars and seasons. In general, berries of seedless cultivars are less well attached to the capstem than seeded cultivars. Shatter symptoms appear at harvest and continue to develop in stored grapes until final consumption. Shatter incidence can be reduced by: (i) optimizing plant growth regulator (PGR) applications during berry development; (ii) controlling pack depth and packing density; (iii) using cluster bagging; (iv) practicing gentle handling and loading; and (v) maintaining the recommended temperature and RH during storage.

Waterberry

This disorder also affects wine and raisin grapes and is known as "bunch-stem necrosis" in Australia, "waterberry" (Fig. 7.11) in Chile, "shanking" in New Zealand, "Stiellahme" in Germany, and "dessechement ed al rafle"

in France. Severely affected vines can have nearly 100% of clusters showing some symptoms and a 50% crop loss in harvested grapes that make
the commercial box (Christensen, 1982). More commonly, crop losses are
5–20% in affected vineyards. Labor to trim affected berries from clusters is
an additional cost factor. The earliest symptom is the development of small
(1–2 mm) dark spots on the capstems (pedicles) and/or other parts of the
rachis framework that most often develops shortly after veraison (berry softening). These spots become necrotic and slightly sunken, then expand
to affect more areas. The affected berries become watery, soft, and flabby,
with an opaque, lighter color when ripe. By harvest, necrosis of the cluster
and capstem phloem is visible. In California, this disorder is associated with
high nitrogen status and vine ammonium concentration, canopy shading,
or cool weather during veraison and ripening (Christensen and Boggero,
1985). 'Thompson Seedless', 'Flame Seedless', and 'Crimson Seedless' are
particularly susceptible to waterberry. Overfertilization with nitrogen and
foliar nutrient sprays of nitrogen should be avoided in waterberry-prone
vineyards. Thinning shoots during ripening can improve light into the
canopy and reduce the problem. Removing affected berries during harvest and packing is a common, though labor-intensive, practice to protect
quality.

Hairline cracks

Sulfur dioxide (SO_2) phytotoxicity due to overexposure causes development of small fine, longitudinal, linear cracks that are almost undetectable to the naked eye (Fig. 7.12). Juice exuding from the split zone makes

Fig. 7.12. Hairline crack symptoms due to overexposure to sulfur dioxide. Photo
courtesy of Dr. Carlos H. Crisosto.

the berry skin wet and sticky. Hairline cracks are observed frequently in clusters packed in boxes using an SO_2-generating pad or when grapes are fumigated excessively prior to precooling (Thompson and Crisosto, 2000; Zoffoli *et al.*, 2008). The incidence of hairline cracks increases when the concentration of SO_2 in a container using SO_2-generating pads exceeds 3 ml l^{-1} h^{-1} SO_2. No hairline cracks are observed at <0.8 ml l^{-1} h^{-1}. Variation among vineyards has been reported and overuse of GA_3 and cytokinin (i.e. CPPU) for berry enlargement contributes to this risk factor. To reduce this disorder, only the minimal dose of SO_2 required for adequate decay protection should be applied.

Berry splitting

Fruit splitting is a common disorder in grapes and includes circumferential or longitudinal injuries, or both, to the grape berry surface. In 'Thompson Seedless', circumferential ring fractures usually appear around the pedicel and longitudinal fractures appear on the sides of overly mature berries. Splitting develops during cell enlargement, coinciding with a high internal hydrostatic pressure (i.e. in 'Flame Seedless'). Turgor pressures of 1.52 MPa and 4.05 MPa are required for splitting in susceptible and resistant grape cultivars, respectively. Splitting can be induced by irrigation or rain and can also occur during storage.

Skin and/or flesh browning

Table grapes commonly suffer from tissue browning during harvest, packing, storage, and shelf life, resulting in lower prices and reduced access to markets. Berry browning is a physiological disorder, primarily of Muscat-type cultivars with high polyphenol concentrations and polyphenol oxidase (PPO) activity, such as 'Italia', 'Regal Seedless', 'BRS Vitória', 'Princess', and 'Majestic' (Fig. 7.2). Symptoms are external (skin) and/or internal (flesh) brown discoloration caused by polyphenol oxidation reactions in the presence of phenolics and PPO in damaged tissues, but non-enzymatic oxidation may also contribute. Skin browning is associated with physical, biological, or chemical damage. For instance, rough handling of clusters during harvest, packaging, or transport often induces skin browning. The incidence of skin browning is low at harvest, but increases after cool storage, particularly in fruit overexposed to sun. Skin browning varies from year to year and among orchards. In 'Princess' table grapes, berries had high skin browning incidence, but very low flesh browning incidence. The most skin browning was found in highly mature grapes and appeared after 3 weeks of cold storage. Skin browning was associated with fruit maturity, but vineyard location had more impact on the incidence of skin browning than maturity. At all locations, the skin browning susceptibility of 'Princess' table grapes increased rapidly when the berries reached a TA of ≤0.60%

Fig. 7.13. Sulfur dioxide berry damage. Photo courtesy of Dr. Carlos H. Crisosto.

and/or a TSS ≥18.0% (Vial *et al.*, 2005; Lichter, 2016). Based on this work, it was recommended to harvest 'Princess' at 16.0–18.0% TSS. Another form of skin browning, termed streak browning and associated with micro cracks, occurs in 'Thompson Seedless' clusters with firm berries that receive excessive pesticide sprays. Methyl bromide phytotoxicity can trigger flesh browning symptoms that affect the entire berry (skin and flesh). Flesh browning symptoms restricted to the vascular bundles or to the endocarp at the seed (or rudimentary seed trace in 'Thompson Seedless') may be related to vineyard and harvesting management. Flesh browning can be also due to chilling injury as discussed above.

Bleaching

Bleaching caused by SO_2 is characterized by partial or total berry discoloration affecting the anthocyanins and chlorophyll, with very well-defined margins (Luvisi *et al.*, 1992). Bleaching usually starts at the pedicel end because of injuries and weak insertion that allows penetration of SO_2 inside the berry base (Fig. 7.13). Often this symptom is known as sunken cap. In some cases, pitting damage has been observed on berries and rachis (Fig. 7.14).

Fig. 7.14. Rachis damage due to overexposure to sulfur dioxide. Photo courtesy of Dr. Carlos H. Crisosto.

Berry Pathology Problems

Gray mold (*Botrytis* rot)

Gray mold (caused by *Botrytis cinerea*) is the most destructive postharvest disease of table grapes, primarily because it develops at temperatures as low as $-0.5°C$ and can spread from berry to berry (Luvisi *et al.*, 1995; Lee *et al.*, 2015; Lichter, 2016). Wounds during harvest provide opportunities for infection, but no wound is required under wet conditions. *Botrytis* rot on grapes can be diagnosed by its characteristic 'slipskin' that develops on the surface of infected berries. Areas infected with gray mold on the berry skin turn brown and slip freely when rubbed, leaving the firm, underlying flesh exposed (Fig. 7.15). Later, white, thread-like hyphal filaments erupt through the berry surface and finally, masses of gray-colored conidia develop. Uncontrolled infections spawn aerial mycelium that spreads to adjacent berries ('nests'). Removing desiccated, infected grapes from the previous season can reduce gray mold infection pressure for the following season. Leaf removal, canopy management, preharvest fungicides, and trimming visibly infected, split, cracked, or otherwise damaged grapes before packing is recommended. Prompt cooling and pre-storage fumigation with sulfur dioxide (discussed below) are used to control gray mold and

Fig. 7.15. A case of high losses due to gray mold (caused by *Botrytis cinerea*) incidence. Photo courtesy of Dr. Carlos H. Crisosto.

sulfur dioxide-generating pads are used during long-distance export to protect grapes during transport or extended storage periods (Palou *et al.*, 2002; Youssef *et al.*, 2015; Chen *et al.*, 2016; Ahmed *et al.*, 2018; Sortino *et al.*, 2018).

Other pathogens (e.g. black rot caused by *Aspergillus* spp., blue rot caused by *Penicillium* spp., and *Rhizopus* rot caused by *Rhizopus* spp. – all discussed below) particularly affect vineyards exposed to warmer temperatures and commonly appear during transport or marketing, after grapes are removed from cold storage. These pathogens become of economic importance in some regions in years with warm, wet springs. Some of these pathogens may need higher concentrations of sulfur dioxide than *Botrytis* for good control.

Cladosporium rot (brown spot)

The prevalence and severity of brown spot on 'Red Globe', 'Crimson Seedless', and other newly released cultivars subjected to long storage periods or long-distance export causes severe losses for Californian and other grape industries. In California, brown spot is caused by three co-occurring *Cladosporium* species: *Cladosporium ramotenellum, Cladosporium cladosporioides* and *Cladosporium limoniforme.* Infection occurs on berries in the field and emerges and spreads from berry to berry after long periods (>60 days) of cold storage. Infection can spread easily from berry to berry through the epidermis, with no wounding necessary, in 'Red Globe' and 'Crimson Seedless' grape berries at temperatures as low as –2°C, potentially causing economic losses. The current practice of storing table grapes at temperatures near 0°C with SO_2 applied at 100–150 ppm h^{-1} can explain why brown rot is becoming a problem during long-term storage and transportation. Using 100–150 ppm SO_2 h^{-1} during cold storage is effective to control gray mold, but does not always prevent brown spot, as *C. cladosporioides* is not fully controlled by this SO_2 concentration. As *C. cladosporioides* is less sensitive to SO_2 applied at 100 ppm h^{-1} than the other pathogenic *Cladosporium* spp. tested, its recovery from brown spot-symptomatic berries may be more frequent. This may explain why *C. cladosporioides* was considered the primary causal agent of brown spot and why 100 ppm SO_2 h^{-1} at common commercial storage temperatures is not always effective against brown spot. A dose of 100 ppm SO_2 h^{-1} did not prevent disease or fungal growth at any temperature, while 400 ppm SO_2 h^{-1} significantly reduced disease incidence and slowed growth of all *Cladosporium* spp. tested at all temperatures. SO_2 applied at 200 or 400 ppm h^{-1} can slow or kill brown spot *Cladosporium* spp., but this treatment is most effective combined with low storage temperatures. Thus, a single application of 200 or 400 ppm SO_2 h^{-1} will essentially eliminate infections through the epidermis on table grapes for at least 30 days when stored or transported near –1°C. Increasing concentrations and efficiency of postharvest SO_2 applications on cold-stored grapes without causing SO_2 bleaching of berries could be a viable management strategy to control postharvest losses from brown spot.

Aspergillus rot

Aspergillus rot (caused by *Aspergillus carbonarius* or *Aspergillus niger*) produces water-soaked brown lesions on wounded berries, followed by dark or black sporulation (conidia). It is primarily associated with warm weather conditions in the vineyard and can be arrested totally, if grapes are cooled rapidly and kept below 4°C. *Aspergillus* rot is often associated with sour rot organisms (yeasts or acetic acid bacteria).

Blue mold

Blue mold decay (caused by *Penicillium expansum*) is distributed worldwide, causing soft rot on grapes, apples, pears, and other fruits during cold storage. It is characterized by light-brown skin discoloration followed by a soft, wet rot that can rapidly affect the entire berry. On the surface of rotten berries, blue-green mold colonies with or without mycelium may appear. An internal watery breakdown often occurs in grapes stored for several weeks at 0°C. Mycelial growth of *P. expansum* can be slowed, but not totally inhibited, at 0°C. Blue mold infections can be initiated in the vineyard, at the packinghouse, or during storage and are associated with wounded fruits, lack of sanitation, and poor management of storage temperature. Conidia are disseminated by wind and contact of infected grapes with healthy ones spreads blue molds efficiently, forming a nest of moldy grapes.

Rhizopus rot

Rhizopus rot (caused by *Rhizopus stolonifer*, also known as *R. nigricans*) usually starts at the base of mature berries as a soft, very watery rot that partially or completely decays infected berries. Longitudinal fissures are produced and black mold develops along them. The skin of the berry turns light gray. Infection is always associated with berries injured during harvest under warm weather conditions. *Rhizopus* rot can be suppressed totally if grapes are stored below 4°C.

Fumigation with Sulfur Dioxide (SO_2)

Fumigation with sulfur dioxide (SO_2) is used to control gray mold because it is not inhibited sufficiently by rapid cooling and cold storage. Standard practice is to fumigate with SO_2 immediately after harvest and/or packing, followed by weekly, lower-dose SO_2 treatments during storage. An exception can be made for grapes produced very early and with an assured rapid sale. For long-distance transport to export markets, SO_2-generating pads are used. Sodium metabisulfite is incorporated into the pads and its reaction with moisture produces a slow release of SO_2 during transit and marketing.

Grape berries and rachis may be damaged by SO_2 fumigation. Symptoms include color bleaching and hairline cracks, followed by sunken areas where accelerated water loss has occurred. These injuries first appear on the berry where some other injury has occurred, such as a harvest wound, transit injury, or breakage at the capstem attachment. Another problem with SO_2 fumigation is the level of sulfite residue remaining on the berries at time of consumption. Sulfur dioxide was once included on the "generally recognized as safe" (GRAS) list of chemicals, for which no registration is required. Heavy usage of sulfites in some other foods has caused a change in regulation because some people are highly allergic to sulfites. Sulfite residues in grapes are currently limited to <10 ppm by the US Environmental Protection Agency and, depending on cultivar, there are limits on the number of repeat SO_2 fumigations allowed. In some countries, there is no maximum residue limit (MRL) for sulfite in grapes; thus, SO_2-treated grapes cannot be shipped to those countries. Also, grapes with residues ≥10 ppm will be rejected. At present, SO_2 is considered a food additive in Europe and requires specific labeling. In the USA, SO_2 is no longer permitted on certified organic grapes, some regulatory agencies do not allow discharge of SO_2 into the air after fumigation, and workers must not be exposed to gas concentrations above 2 mg l[-1].

Careful attention to SO_2 treatment procedures is necessary to minimize damage. Because of potential SO_2 problems, California has used a total utilization technique for the last 12 years. This technique differs from the traditional system in that there is no excess SO_2 fumigant at the end of treatment, reducing air pollution, sulfite residues in the fruit, and potential worker exposure. This SO_2 application is based on the premise that at least 100 ppm SO_2 h[-1] (100 concentration × time product (CT)) was necessary to kill *B. cinerea* conidia and inactivate exposed mycelia at 0°C. Even less than 100 CT was required on warm grapes to control conidia and mycelium. The sensitivity of *B. cinerea* conidia to SO_2 increases two- to fourfold for every 10°C increment between 0°C and 32°C, because of the temperature effect on SO_2 absorption. SO_2 applied at 200 ppm 30 min[-1] or 400 ppm 15 min[-1] was as effective as 100 ppm h[-1]. The other requirement is to apply the initial SO_2 (~100–150 ppm h[-1]) during the initial forced-air cooling after harvest, with subsequent weekly applications during cold storage with good air distribution. Total utilization typically uses ~10 times less SO_2 compared with traditional systems, but requires uniform distribution of air in the room for effective treatment. SO_2 does not penetrate the intact berry skin, remaining almost completely on the berry surface. Therefore, it kills spores and mycelia present on the berry surface, but cannot control the fungus inside grapes or underneath the skin. However, SO_2 is absorbed efficiently through berry injuries such as hairline cracks, splitting, and non-suberized lenticels, resulting in partial or entire berry bleaching. The fungal toxicity of SO_2 treatment is attributed almost entirely to the fact that SO_2 is absorbed passively through the plasma membrane, causing oxidation reactions that affect different metabolic processes of *B. cinerea* and other fungi.

Initial fumigation

The first SO_2 fumigation after harvest is carried out in a sealed room with controlled venting and exhaust fans injecting SO_2 from a compressed cylinder (Fig. 7.16). In most operations, this fumigation is applied concurrent with forced-air cooling to ensure even penetration of SO_2 to the center boxes within a pallet. In some cases, an initial sulfur dioxide fumigation can occur on harvested grapes prior to packing, which is typical in Chile, by using small fumigation chambers (Fig. 7.17). In most combinations of boxes and packs, this system produces >85% penetration, measured as the percentage of the room air CT product. SO_2 at 0.1% penetrated grapes packed in perforated box liners or clamshells very well. In other production areas, a gun that injects SO_2 into each box is being used with mixed results.

Fig. 7.16. Sulfur dioxide applied from a compressed gas cylinder during an initial fumigation in California. Photo courtesy of Dr. Carlos H. Crisosto.

Weekly cold-storage fumigation

This fumigation process is applied every 7–10 days. After SO_2 application in the storage room, fans in the room should run at high speed to maintain airflow of $0.05\,m^3\,t^{-1}$ for >3 h so that the fruit, packaging materials, or room surfaces absorb nearly all of the SO_2. At the end of fumigation, the concentration of SO_2 in the room air should be <2–5 ppm and no venting or scrubbing should be needed. In this common system, each cold storage room should be calibrated to determine the amount of SO_2 to use. Center boxes within a pallet have less SO_2 exposure than corner boxes and pallets closest to the SO_2 inlet have greater exposure than those farthest away. There is a large difference in SO_2 penetration of box materials: for example, SO_2 penetration is best in EPS boxes, moderate in wood-end boxes, and lowest in corrugated boxes. Fumigant penetration and distribution should be checked with SO_2 dosimeter tubes (Fig. 7.18). Dosimeter tubes should be

Fig. 7.17. Sulfur dioxide fumigation chamber used in the shed-packing handling system. Photo courtesy of Dr. Carlos H. Crisosto.

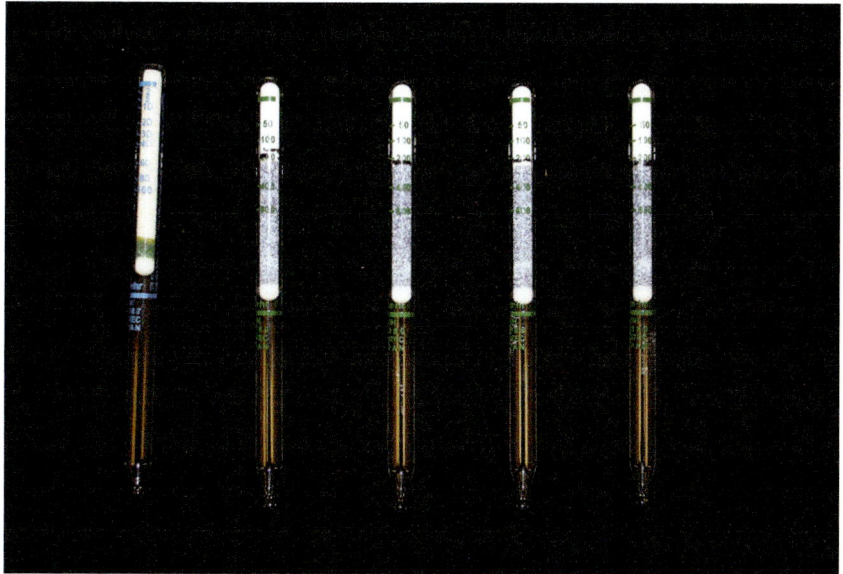

Fig. 7.18. Dosimeters used to measure sulfur dioxide applied during fumigation. The last four dosimeters on the right-hand side show 200 ppm h^{-1}. Photo courtesy of Dr. Carlos H. Crisosto.

placed in the center of the boxes inside tissue wraps of cluster bags if these are present, and usually in boxes located in the center of the pallets. The tubes should be examined promptly after fumigation since the color reaction may continue and lead to an overestimate of the dosage. This analysis allows the operator to adjust the amount of fumigant applied to ensure that grapes in most boxes are adequately protected from decay (a minimum of 100 ppm h^{-1}), but not exposed to fumigant concentrations that might cause excessive residues and bleaching. Grapes should be monitored regularly during storage for physiological deterioration, fruit rot, SO_2 injury, and rachis browning-drying. Grapes should receive an additional SO_2 fumigation before loading for transport/shipment to assure a longer market life because fumigation is seldom available in receiving markets.

A special, very low SO_2 dosage has been developed for fumigation in trucks and overseas containers (Crisosto *et al.*, 2002c). Unless SO_2 fumigation is available, the receiver must order grapes for immediate needs and must complete distribution and marketing within a reasonable time after arrival. An exception would be when SO_2-generating pads are placed in the container before shipment. Washing marine containers with sodium hydroxide will help prevent container damage due to SO_2.

Long-distance shipment

For long-distance shipments involving journeys lasting 10 days or longer, during which usual SO_2 fumigation cannot be applied, SO_2-generating pads in combination with a box liner and packaging materials is advised.

The Box Liner and SO_2-Generating Pad System

The fumigation system consisting of a box liner combined with a sodium metabisulfite-embedded pad involves other inner packaging components to make the system more effective and less risky. These components must act in balance for a safe and effective performance: (i) SO_2-generating pad (Fig. 7.19); (ii) container; (iii) box liner (Fig. 7.20); (iv) cluster bag (Fig. 7.21); (v) bottom pad; and (vi) other inner paper packaging. The SO_2-generating pad replaces periodic SO_2 fumigations for gray mold control during shipment, in combination with polyethylene box liners and low temperature. These SO_2-generating pads were first developed at UC Davis in the late 1960s and are now used worldwide, especially in production areas with large export markets such as California, Chile, South Africa, Australia, Italy, Greece, and Egypt. This system depends on gaseous sulfur dioxide released after reaction of sodium metabisulfite with environmental moisture. A portion of this gas will escape out of the box, especially in packages without liners or with highly perforated liners, and more will be absorbed by the packaging materials before reaching the air spaces surrounding the

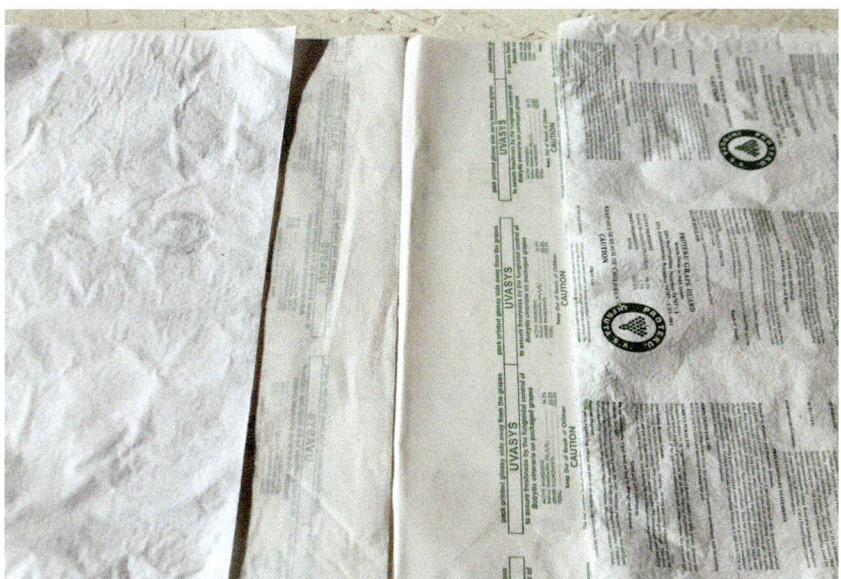

Fig. 7.19. Sulfur dioxide-generating pads. Photo courtesy of Dr. Carlos H. Crisosto.

Fig. 7.20. Perforated box liner which is used in combination with a sulfur dioxide-generating pad. Photo courtesy of Dr. Carlos H. Crisosto.

berries; however, a portion of the gas will reach those air spaces and come into contact with the grapes. Sulfur dioxide can react with surface materials on the skin, including fungal structures (surface-borne spores and growing mycelia), or penetrate through natural openings or wounds, where it reacts

Fig. 7.21. Restricted cluster bag (RCB). Photo courtesy of Dr. Carlos H. Crisosto.

with constituents of the epidermal cells or is diluted in the tissue liquids. Depending on the gas concentration, system balance, and other factors, consequences such as fungal inhibition, phytotoxicity, and/or accumulation of sulfite residues can occur. SO_2-generating pads have historically been made from paper and recently from plastic (polyethylene) materials of different dimensions. In both types, the sodium metabisulfite ($Na_2S_2O_5$), when hydrated by water vapor, continuously emits SO_2, protecting grapes from decay for up to 60 days during cold storage. The rate of emission of SO_2 from a generator pad is proportional to the temperature and humidity inside the box, the material type, and the manufacturing design. In general, one paper pad containing 8 g $Na_2S_2O_5$ in an 8.2 kg box is placed over the grapes in each box, where ~0.1 g $Na_2S_2O_5$ day^{-1} is released.

There are two types of SO_2 pad emission: (i) dual release (fast, then slow); and (ii) slow release. The different emission rates are attained in different ways, such as combining paper and plastic, or plastic laminated and perforated with different size microperforations. In most dual formulations, the SO_2-generating pads contain 1 g or 7 g of $Na_2S_2O_5$ for the fast- and slow-release formulations, respectively. In Chile and South Africa, dual-release generator pads are required. The fast phase is activated and depleted during forced-air cooling. Therefore, the SO_2 concentration during cold storage and/or transportation depends on the slow-release phase, which is liberated slowly or rapidly depending on temperature variation, RH, condensation inside the box, and pad design. In California, a slow-release SO_2-generating pad, combined with a perforated polyethylene box liner with holes of 6.4 mm spaced every 7.6 cm (0.3% perforation) or spaced every 10.2 cm (0.6% perforation), reduces water loss and assures gray mold control without enhancing SO_2 phytotoxicity. There are new products that combine a box liner with an SO_2 generator, where the liner contains in its structure sodium metabisulfite, which in contact with moisture generates sulfur dioxide.

The box liner or perforated bag

Our previous work demonstrated that for Californian handling conditions, the best packaging option for exported table grapes was a strong polyethylene box liner with a 0.3–1.2% vented area (VA) (Fig. 7.20), combined with a slow-release sulfur dioxide-generating pad, after an initial sulfur dioxide fumigation. These VAs were reached by using microperforated (micro-punctures or needle size) and perforated polyethylene box liners with 0.64 cm-diameter holes spaced every 7.6 cm or every 10.2 cm. Solid and poorly perforated polyethylene box liners have been used in the South African and Chilean industries, respectively, for a long time. Recently, due to the United States Department of Agriculture (USDA) Animal and Plant Health Inspection Service (APHIS), Chileans were forced to change their box liner VA from 0.3% to 0.9% to 1.2%. These regulation changes require perforations of 0.476 cm or 0.635 cm diameter and spaced at 4.4 cm or 5 cm, respectively. This new hole size and distribution results in a 0.9–1.2% VA. Previously, the VA of the commercial box liner was 0.3% (0.64 cm holes, spaced every 7.6 cm). This increased VA affects their box-liner-system performance because the other components of the box liner system were not adjusted. The use of the box liner increases potential condensation and cooling time, and reduces weight loss and SO_2 penetration.

The bottom pad

As a component of the box liner system, the bottom pad has the primary function of absorbing water that is deposited on the bottom of the box as a result of temperature changes during handling. Therefore, this pad must have a water absorption capacity and ideally cover most of the box bottom without blocking air channels. A bottom pad also reduces berry browning on green colored cultivars due to abrasion or compression damage against the bottom of the box, the critical function of this pad as a water absorbent has been forgotten and plastic pad material without water absorption capacity is used commercially. In a field test, a bottom pad reduced the number of bruised berries by nearly one half. Bruising incidence was almost double in boxes toward the bottom of the pallet than in those toward the top of the pallet. This suggests that bruising damage is associated with pallet loading. Thus, a large reduction in the number of bruised berries can be achieved with a one-layer box, bottom pad, and gentle handling during pallet loading. Some SO_2-generating bottom pads have been used to improve SO_2 distribution in the box and control potential decay, but the potential benefits are outweighed by high bleaching incidence.

Inner paper packaging

As they can absorb SO_2, paper wraps or other paper packaging materials interfere with the availability of SO_2. Some of the gas released by the pad will escape out of the box, especially in packages without liners or with highly perforated liners. Another fraction will be absorbed by the packaging materials before reaching the air spaces surrounding the berries. However, the most important portion of the gas will reach those air spaces and come into contact with the grapes. Sulfur dioxide can react with surface materials on the skin, including fungal structures (surface-borne spores and growing mycelia), or penetrate through natural openings or wounds to react with constituents of the epidermal cells or be diluted in the tissue liquids. All these processes count as absorption of the gas by the grapes. Depending on gas concentration and other factors, consequences such as fungal inhibition, phytotoxicity, and/or accumulation of sulfite residues can occur.

Restricted cluster bag (RCB)

An RCB (1.4% VA) slows drying and shriveling of fruit and rachis by limiting water loss. RCB and standard bags were tested in several cultivars using the commercial cluster bag (with 60% perforation) or RCB (with 1.4% perforation) in foam boxes (Fig. 7.21). The RCB reduced rachis browning and increased the buyer opinion grade without affecting decay and phytotoxicity under standard Californian conditions. Grapes packed in bags with 1.4% perforation were categorized as 'good' according to the buyer opinion grade. There was less shattering in grapes packed in the RCB than in grapes packed with the commercial bag: 16.7% instead of 21.3%. SO_2 penetration was adequate in both bag types during the initial treatment and weekly fumigations. The RCB more effectively reduced water loss and maintained rachis freshness without interfering with SO_2 penetration compared with the standard commercial cluster bag. During storage, neither the RCB nor the commercial cluster bag showed excessive condensation. However, grapes packed in RCB with high restriction to VAs had condensation, decay, and bleaching problems during these evaluations.

New Alternatives to SO_2

There has been interest in replacing and/or complementing SO_2 by using biological control, physical methods, or GRAS chemicals. Heat, ethanol (Venditti *et al.*, 2017), edible coatings (Chen *et al.*, 2019), chitosan (Shen and Yang, 2017), carbon dioxide (Teles *et al.*, 2014; Cefola and Pace, 2016; Silva-Sanzana *et al.*, 2016; Altieri *et al.*, 2018), chlorine dioxide (Lichter, 2016), ozone (Admane *et al.*, 2018), acetic acid vapor, chlorine pads, pulsated ultraviolet, ultraviolet (Freitas *et al.*, 2015), *Muscodor* volatile fumigant (Gabler *et al.*, 2010a), ozone (Gabler *et al.*, 2010b), and carbonates have been tested. Unfortunately, among these methods, only the combinations of 10% or 15% CO_2 with 3%, 6%, or 12% O_2 were effective in controlling *Botrytis*. The best and safest control of *B. cinerea* development occurred with 15 kPa CO_2, but high CO_2 concentrations can cause damage in some cultivars, depending on storage time and maturity. The specific relationships between controlled atmosphere (CA) conditions, maturity, and decay control in green, black, and red table grapes was tested in California (Crisosto *et al.*, 2002a, b). A dose of 10–15 kPa CO_2 is recommended for up to 12 weeks storage of late harvested 'Thompson Seedless' table grapes (19.0% TSS). CA is not recommended for early harvested (16.5% TSS) 'Thompson Seedless' grapes. The main storage limitations for early-harvested grapes were development of 'off-flavors' and rachis and berry browning resulting from exposure to >10 kPa CO_2. For 'Red Globe', 10 kPa CO_2 is suggested for up to 12 weeks storage of late-harvested grapes. For early-harvested 'Red Globe' grapes, an atmosphere of 10 kPa CO_2 is suggested for periods shorter than 4 weeks. Off-flavors developed in 'Superior' grapes stored for 6 weeks in 15 kPa CO_2, but not in 10 kPa CO_2, and 5 kPa O_2. In these evaluations, all fruit were fumigated initially with SO_2 and air-stored grapes were used as controls. Natural quiescent *Botrytis* infection ranged from none to 5.5% in the first month for air-stored, early-harvested grapes.

For late-harvested grapes, natural *Botrytis* infection varied from 1.0% to 32.8%. In conditions of high inoculum (simulating bad years), an initial SO_2 fumigation was necessary to keep decay development low during postharvest life.

Modified atmosphere packaging (MAP) has not spread to commercial use because the non-perforated MAP liner increases cooling times and interferes with the fast, even-cooling operation demanded by current table grape handling dogma. In addition, most MAP liners do not reach the 10–15% CO_2 at low temperatures required to control *Botrytis* rot development and the increased condensation inside the package can activate *Botrytis* penetration into berries and increase decay losses. In some instances, MAP makes it difficult to implement quarantine treatments required by some markets.

In general, these alternatives have a direct effect on control of *Botrytis* and an indirect effect on increasing volatiles and fruit natural resistances (Lichter *et al.*, 2006; Cefola *et al.*, 2018). Lack of contact between chemicals and organisms, potential damage, and treatment persistence form the main commercial limitations. This area of research is very active, including studies using combinations of treatments and application times, pre- and postharvest, to complement action, increase persistence, and reduce potential environmental issues.

Suitability as a Fresh-cut Product

Berries are highly suitable for the fresh-cut market, although some cultivars with clear abscission layers fit this category better than others. Equipment for removing berries from the rachis and methods for proper sanitation and packaging are commercially available.

Utilization

Table grapes are mainly consumed fresh, but fruit that does not make the pack can be used as grape juice, partially fermented grape juice ("chicha" in some Latin-American countries), wine, alcoholic beverages (brandy, cognac, grappa, jerez, oporto, pisco, and singani), wine vinegar, grape jam, grape preserves, and raisins. Additionally, grapes can be used to produce grape seed oil used for cooking, salad dressings, or cosmetic products such as body creams and lip balm. Although specific grape cultivars are used to elaborate each product, table grape cultivars are often produced both for fresh consumption and for drying as raisins. One niche market has been created in Spain to fulfill their old good luck tradition during New Year's Eve. For this, they consume a berry per minute starting 12 min before midnight to say goodbye to the Old Year and welcome in the New Year. In order to satisfy this limited demand, clusters from a special late cultivar are paper bagged during growth and development to protect the berries

Fig. 7.22. Bagged clusters to protect grapes for late harvest in Spain. Photo courtesy of Dr. Carlos H. Crisosto.

and table grapes that are harvested in late November and early December (Fig. 7.22).

Special and Quarantine Treatments

Table grapes for the export market must comply with sanitary requirements imposed by the importing countries. Rules for import into the USA are issued by the USDA APHIS. This agency also provides information to assist exporters in targeting markets and defining what entry requirements a country might have for US table grapes. APHIS also provides phytosanitary inspections and certifications that grapes are free of pests to facilitate compliance with foreign regulatory requirements. Of primary concern are the vine moth, *Lobesia botrana,* the Mediterranean fruit fly, *Ceratitis capitata,* and miscellaneous external-feeding insects. Frequently, grapes imported into the USA from compromised countries are fumigated with methyl bromide, following treatment schedules issued by APHIS, to prevent entry of insect pests. Cold treatments are also accepted by APHIS for control of fruit flies. Grapes exported from the USA may harbor pests of concern elsewhere, but rarely require treatment, although this situation can change rapidly. For example, black widow spiders are occasional hitchhikers within grape clusters or within grape boxes. An official approved protocol based on CO_2 and SO_2 fumigation is used in California to kill black widow spiders in packaged table grapes. This protocol consists of 30 min fumigation with 6% CO_2 and

1% SO_2. The design and materials of inner packaging and containers can influence the concentration of SO_2 surrounding grapes inside the containers. Because container types affect SO_2 penetration and thus the success of the fumigation, they must be approved by the USDA.

Special Research Needs

Currently, the most important research topic is the replacement of SO_2 with less toxic chemicals and/or organic approaches. Introduction of table grape cultivars with predominant flavors and different shapes is also an interesting research area.

Acknowledgments

University of California Agriculture and Natural Resources (UC ANR) retains the copyright of the short version of this table grape chapter that will be published in *Postharvest Technology of Horticulture Crops*, 4th edition and CABI will hold the copyright for this long version.

References

Admane, N., Genovese, F., Altieri, G., Tauriello, A., Trani, A. *et al.* (2018) Effect of ozone or carbon dioxide pre-treatment during long-term storage of organic table grapes with modified atmosphere packaging. *LWT– Food Science and Technology* 98, 170–178. DOI: 10.1016/j.lwt.2018.08.041.

Ahmed, S., Roberto, S., Domingues, A., Shahab, M., Junior, O. *et al.* (2018) Effects of different sulfur dioxide pads on *Botrytis* mold in 'Italia' table grapes under cold storage. *Horticulturae* 4(4), 29. DOI: 10.3390/horticulturae4040029.

Altieri, G., Genovese, F., Matera, A., Tauriello, A. and Di Renzo, G.C. (2018) Characterization of an innovative device controlling gaseous exchange in packages for food products. *Postharvest Biology and Technology* 138, 64–73. DOI: 10.1016/j.postharvbio.2017.12.012.

Avenant, J.H. (2017) Effect of gibberellic acid, CPPU and harvest time on browning of *Vitis vinifera* L. 'Regal Seedless': pre-cold storage and post-cold storage quality factors. *Acta Horticulturae* 1157, 373–380.

Bahar, A., Kaplunov, T., Alchanatis, V. and Lichter, A. (2017) Evaluation of methods for determining rachis browning in table grapes. *Postharvest Biology and Technology* 134, 106–113. DOI: 10.1016/j.postharvbio.2017.08.016.

Blanckenberg, A., Fawole, O.A. and Opara, U.L. (2018) Quantifying postharvest losses of 'Crimson Seedless' table grapes along the supply chain. *Acta Horticulturae* 1201, 29–34. DOI: 10.17660/ActaHortic.2018.1201.5.

Cantín, C.M., Fidelibus, M.W. and Crisosto, C.H. (2007) Application of abscisic acid (ABA) at veraison advanced red color development and maintained postharvest quality of 'Crimson Seedless' grapes. *Postharvest Biology and Technology* 46(3), 237–241. DOI: 10.1016/j.postharvbio.2007.05.017.

Cappellini, R.A., Ceponis, M.J. and Lightner, G.W. (1986) Disorders in table grape shipments to the New York market, 1972–1984. *Plant Disease* 70(11), 1075–1079. DOI: 10.1094/PD-70-1075.

Cefola, M. and Pace, B. (2016) High CO_2-modified atmosphere to preserve sensory and nutritional quality of organic table grape (cv. 'Italia') during storage and shelf-life. *European Journal of Horticultural Science* 81(4), 197–203. DOI: 10.17660/eJHS.2016/81.4.2.

Cefola, M., Damascelli, A., Lippolis, V., Cervellieri, S., Linsalata, V. *et al.* (2018) Relationships among volatile metabolites, quality and sensory parameters of 'Italia' table grapes assessed during cold storage in low or high CO_2 modified atmospheres. *Postharvest Biology and Technology* 142, 124–134. DOI: 10.1016/j.postharvbio.2017.09.002.

Chen, R., Wu, P., Cao, D., Tian, H., Chen, C. *et al.* (2019) Edible coatings inhibit the postharvest berry abscission of table grapes caused by sulfur dioxide during storage. *Postharvest Biology and Technology* 152, 1–8. DOI: 10.1016/j.postharvbio.2019.02.012.

Chen, X., Mu, W., Peter, S., Zhang, X. and Zhu, Z. (2016) The effects of constant concentrations of sulfur dioxide on the quality evolution of postharvest table grapes. *Journal of Food and Nutrition Research* 55, 114–120.

Christensen, L.P. (1982) Waterberry – what we know today. In: *Table Grape Seminar Proceedings*. University of California Cooperative Extension and California Table Grape Commission, Dinuba, California, pp. 12–14.

Christensen, L.P. and Boggero, J.D. (1985) A study of mineral nutrition relationships of waterberry in 'Thompson Seedless'. *American Journal of Enology and Viticulture* 36, 57–64.

Crisosto, C.H. and Crisosto, G.M. (2002) Understanding American and Chinese consumer acceptance of 'Redglobe' table grapes. *Postharvest Biology and Technology* 24(2), 155–162. DOI: 10.1016/S0925-5214(01)00189-2.

Crisosto, C.H., Smilanick, J.L. and Dokoozlian, N.K. (2001) Table grapes suffer water loss, stem browning during cooling delays. *California Agriculture* 55(1), 39–42. DOI: 10.3733/ca.v055n01p39.

Crisosto, C.H., Garner, D. and Crisosto, G. (2002a) Carbon dioxide-enriched atmospheres during cold storage limit losses from *Botrytis* but accelerate rachis browning of 'Red Globe' table grapes. *Postharvest Biology and Technology* 26(2), 181–189. DOI: 10.1016/S0925-5214(02)00013-3.

Crisosto, C.H., Garner, D. and Crisosto, G.M. (2002b) High carbon dioxide atmospheres affect stored 'Thompson Seedless' table grapes. *HortScience* 37(7), 1074–1078. DOI: 10.21273/HORTSCI.37.7.1074.

Crisosto, C.H., Palou, L., Garner, D. and Armson, D.A. (2002c) Concentration by time product and gas penetration after marine container fumigation of table grapes with reduced doses of sulfur dioxide. *HortTechnology* 12(2), 241–245. DOI: 10.21273/HORTTECH.12.2.241.

Freitas, P.M., López-Gálvez, F., Tudela, J.A., Gil, M.I. and Allende, A. (2015) Postharvest treatment of table grapes with ultraviolet-C and chitosan coating preserves quality and increases stilbene content. *Postharvest Biology and Technology* 105, 51–57. DOI: 10.1016/j.postharvbio.2015.03.011.

Gabler, F.M., Mercier, J., Jiménez, J.I. and Smilanick, J.L. (2010a) Integration of continuous biofumigation with *Muscodor albus* with pre-cooling fumigation with ozone or sulfur dioxide to control postharvest gray mold of table grapes. *Postharvest Biology and Technology* 55(2), 78–84. DOI: 10.1016/j. postharvbio.2009.07.012.

Gabler, F.M., Smilanick, J.L., Mansour, M.F. and Karaca, H. (2010b) Influence of fumigation with high concentrations of ozone gas on postharvest gray mold and fungicide residues on table grapes. *Postharvest Biology and Technology* 55(2), 85–90. DOI: 10.1016/j.postharvbio.2009.09.004.

International Organisation of Vine and Wine (2019) 2019 Statistical Report on World Vitiviniculture. International Organisation of Vine and Wine, Paris. Available at: http://www.oiv.int/public/medias/6782/oiv-2019-statistical-report-on-world-vitiviniculture.pdf (accessed 31 October 2019).

Lee, J.-S., Kaplunov, T., Zutahy, Y., Daus, A., Alkan, N. *et al.* (2015) The significance of postharvest disinfection for prevention of internal decay of table grapes after storage. *Scientia Horticulturae* 192, 346–349. DOI: 10.1016/j. scienta.2015.06.026.

Li, L., Kaplunov, T., Zutahy, Y., Daus, A., Porat, R. *et al.* (2015) The effects of 1-methylcyclopropane and ethylene on postharvest rachis browning in table grapes. *Postharvest Biology and Technology* 107, 16–22. DOI: 10.1016/j. postharvbio.2015.04.001.

Lichter, A. (2016) Rachis browning in tablegrapes. *Australian Journal of Grape and Wine Research* 22(2), 161–168. DOI: 10.1111/ajgw.12205.

Lichter, A., Gabler, F.M. and Smilanick, J.L. (2006) Control of spoilage in table grapes. *Stewart Postharvest Review* 2, 1–10.

Luvisi, D.A., Shorey, H., Smilanick, J., Thompson, J., Gump, B. *et al.* (1992) *Sulfur dioxide fumigation of table grapes.* University of California DANR Bulletin #1932, University of California DANR, Oakland, California.

Luvisi, D., Shorey, H., Thompson, J., Hinsch, T. and Slaughter, D. (1995) *Packaging California grapes.* University of California DANR Publication #1934, University of California DANR, Oakland, California.

Palou, L., Crisosto, C.H., Garner, D., Basinal, L.M., Smilanick, J.L. *et al.* (2002) Minimum constant sulfur dioxide emission rates to control gray mold of cold-stored table grapes. *American Journal of Enology and Viticulture* 53, 110–115.

Palou, L., Crisosto, C.H., Garner, D. and Basinal, L.M. (2003) Effect of continuous exposure to exogenous ethylene during cold storage on postharvest decay development and quality attributes of stone fruits and table grapes. *Postharvest Biology and Technology* 27(3), 243–254. DOI: 10.1016/ S0925-5214(02)00112-6.

Pilati, S., Bagagli, G., Sonego, P., Moretto, M., Brazzale, D. *et al.* (2017) Abscisic acid is a major regulator of grape berry ripening onset: new insights into ABA signaling network. *Frontiers in Plant Science* 8, 1093. DOI: 10.3389/fpls.2017.01093.

Shen, Y. and Yang, H. (2017) Effect of preharvest chitosan-g-salicylic acid treatment on postharvest table grape quality, shelf life, and resistance to *Botrytis cinerea*-induced spoilage. *Scientia Horticulturae* 224, 367–373. DOI: 10.1016/j. scienta.2017.06.046.

Silva-Sanzana, C., Balic, I., Sepúlveda, P., Olmedo, P., León, G. *et al.* (2016) Effect of modified atmosphere packaging (MAP) on rachis quality of 'Red Globe' table grape variety. *Postharvest Biology and Technology* 119, 33–40. DOI: 10.1016/j. postharvbio.2016.04.021.

Sortino, G., Farina, V., Gallotta, A. and Allegra, A. (2018) Effect of low SO_2 postharvest treatment on quality parameters of 'Italia' table grape during

prolonged cold storage. *Acta Horticulturae* 1194,695–700. DOI: 10.17660/
ActaHortic.2018.1194.99.

Teles, C.S., Benedetti, B.C., Gubler, W.D. and Crisosto, C.H. (2014) Prestorage application of high carbon dioxide combined with controlled atmosphere storage as a dual approach to control *Botrytis cinerea* in organic 'Flame Seedless' and 'Crimson Seedless' table grapes. *Postharvest Biology and Technology* 89, 32–39. DOI: 10.1016/j.postharvbio.2013.11.001.

Thompson, J.F. and Crisosto, C. (2000) Handling at destination markets. In: Kader, A.A. (ed.) *Postharvest Technology of Horticultural Crops.* University of California DANR, Oakland, California, pp. 271–279.

Venditti, T., Ladu, G., Cubaiu, L., Myronycheva, O. and D'hallewin, G. (2017) Repeated treatments with acetic acid vapors during storage preserve table grapes fruit quality. *Postharvest Biology and Technology* 125, 91–98. DOI: 10.1016/j.postharvbio.2016.11.010.

Vial, P.M., Crisosto, C.H. and Crisosto, G.M. (2005) Early harvest delays berry skin browning of 'Princess' table grapes. *California Agriculture* 59(2), 103–108. DOI: 10.3733/ca.v059n02p103.

Wang, Z., Zhou, J., Xu, X., Perl, A., Chen, S. *et al.* (2017) Adoption of table grape cultivars: an attribute preference study on Chinese grape growers. *Scientia Horticulturae* 216, 66–75. DOI: 10.1016/j.scienta.2017.01.001.

Youssef, K., Roberto, S.R., Chiarotti, F., Koyama, R., Hussain, I. *et al.* (2015) Control of *Botrytis* mold of the new seedless grape 'BRS Vitoria' during cold storage. *Scientia Horticulturae* 193, 316–321. DOI: 10.1016/j.scienta.2015.07.026.

Zoffoli, J.P., Latorre, B.A. and Naranjo, P. (2008) Hairline, a postharvest cracking disorder in table grapes induced by sulfur dioxide. *Postharvest Biology and Technology* 47(1), 90–97. DOI: 10.1016/j.postharvbio.2007.06.013.

Appendix: Storage Requirements, and Benefits of Postharvest Treatment for Mediterranean Tree Fruit

| Commodity | Storage requirements | | Main fruit problems | Postharvest treatment benefits[a] |
	Ideal storage temperature (°C)	Relative humidity (%)		
Almond	0–10	The current recommended optimum storage temperature is 0–10°C, and 50–65% relative humidity ('safe moisture level') should be maintained to keep ~2.8–7.0% moisture in the kernels with a water activity (aw) of 0.2–0.8 at 20°C that does not support fungal growth and affect sensory attributes	Aflatoxin, concealed damage, insects, odor transfer	Insect fumigation, drying
Fresh fig	−1–0	85–90	Diseases	Sulfur dioxide, controlled atmosphere
Peach	−1–0	90–95	Chilling injury, brown rot, sour rot, and gray mold	Preconditioning
Persimmon	1–0	90–95	Flesh browning, chilling injury, softening, ethylene sensitivity	Desastringency, 1-MCP, controlled atmosphere
Pistachio	0–10	The current recommended optimum storage temperature is 0–10°C, and 50–65% relative humidity ('safe moisture level') should be maintained to keep ~2.8–7.0% moisture in the kernels with a water activity (aw) of 0.2–0.8 at 20°C that does not support fungal growth and affect sensory attributes	Aflatoxin staining, insects, odor transfer	Drying

© CAB International 2020. *Manual on Postharvest Handling of Mediterranean Tree Fruits and Nuts* (eds Carlos H. Crisosto and Gayle M. Crisosto)

| | Storage requirements | | | |
Commodity	Ideal storage temperature (°C)	Relative humidity (%)	Main fruit problems	Postharvest treatment benefits[a]
Pomegranate	5–7	90–95	Chilling injury, gray mold, heart rot, and cracking	Fungicide immersion, controlled atmosphere
Table grape	−1–0	90–95	Gray mold, rachis browning and physiological disorders	Sulfur dioxide, controlled atmosphere

[a]1-MCP, 1-methylcyclopropene.

Index

Note: Page numbers in **bold** type refer
to **figures**

Page numbers in *italic* type refer
to *tables*

CABI – who we are and what we do

This book is published by **CABI**, an international not-for-profit organisation that improves people's lives worldwide by providing information and applying scientific expertise to solve problems in agriculture and the environment.

CABI is also a global publisher producing key scientific publications, including world renowned databases, as well as compendia, books, ebooks and full text electronic resources. We publish content in a wide range of subject areas including: agriculture and crop science / animal and veterinary sciences / ecology and conservation / environmental science / horticulture and plant sciences / human health, food science and nutrition / international development / leisure and tourism.

The profits from CABI's publishing activities enable us to work with farming communities around the world, supporting them as they battle with poor soil, invasive species and pests and diseases, to improve their livelihoods and help provide food for an ever growing population.

CABI is an international intergovernmental organisation, and we gratefully acknowledge the core financial support from our member countries (and lead agencies) including:

Discover more

To read more about CABI's work, please visit: **www.cabi.org**

Browse our books at: **www.cabi.org/bookshop**,
or explore our online products at: **www.cabi.org/publishing-products**

Interested in writing for CABI? Find our author guidelines here:
www.cabi.org/publishing-products/information-for-authors/